DOWNEAST GENIUS

From Earmuffs to Motor Cars
Maine Inventors Who Changed the World

Other Books by Islandport Press

Forever Yours, Bar Harbor
By Earl Brechlin

Wild! Weird! Wonderful! Maine.
By Earl Brechlin

This Day in Maine
By Joseph Owen

Whatever It Takes
May Davidson

Hauling By Hand
Dean L. Lunt

Here for Generations
Dean L. Lunt

Nine Mile Bridge
Helen Hamlin

Shoutin' into the Fog
Thomas Hanna

These and other Maine books available at www.islandportpress.com

DOWNEAST GENIUS

From Earmuffs to Motor Cars
Maine Inventors Who Changed the World

EARL H. SMITH

ISLANDPORT PRESS

Islandport Press

Islandport Press
P.O. Box 10
Yarmouth, Maine 04096
www.islandportpress.com
info@islandportpress.com

First printing, May 2021

ISBN: 978-1-952143-27-4
Library of Congress Card Number: 2021932657
Printed in USA.

Dean L. Lunt, Publisher
Teresa Lagrange, Book designer

On the cover (clockwise from top left): illustration for train wheel patent, 1836; Chester Greenwood, 1877; illustration for paper bag machine patent; Obed Hussey; illustration for train wheel patent; Helen Blanchard; illustration for earmuff patent, 1877; Hiram Maxim, 1884.

Back cover photo of Earl Smith by Dave Dostie.

For Aidan Smith
an irrepressible tinkerer

Table of Contents

Author's Note

The genius Thomas Edison once said all anyone needed to invent something was a good imagination and a pile of junk. As things turned out, imagination and creativity have always been essential qualities for survival in Maine, and the ingrained habits of "saving up" and "making do" inevitably produce heaps of junk. With Edison's formula in mind, it's not surprising Mainers were quick to join the parade of American inventors who began their march at the beginning of the nineteenth century.

George Washington had seen the coming rush, and in his first State of the Union address in 1790 he called for the establishment of a United States Patent Office. Four years later, Washington signed the license for Eli Whitney's cotton gin and a stampede of remarkable new inventions began.

In 1820, when the District of Maine divorced the Commonwealth of Massachusetts, there were barely three hundred thousand souls ready to manage the vast new state. With twenty-two million acres of land, each person could have had some eighty acres all to themselves. The natural riches were unimaginable. Maine's forests held an endless supply of timber, and the cleared, rich land yielded valuable crops. Maine's abundant waters included the most powerful rivers east of the Mississippi, and a long, winding coastline boasted some of the finest ports on Earth. With all these natural gifts, Maine people found themselves well poised to meet the challenges of the coming Industrial Revolution and its unquenchable thirst for new and better things.

In the aftermath of the Civil War, harnessing electricity and the rapidly expanding railroads provided fuel for even more American inventors, and Maine geniuses were thick among them. Throughout the second half of the century, the United States Patent Office was flooded with more than 250 applications a day, and by 1900, the number of new licenses soared above 640,000.

Beyond having ties to Maine, a common theme amongst all the inventors in this book is that most began as ordinary folks, untrained farmers and laborers, who chased their dreams in cluttered barns and crude workshops from York to Van Buren and from North Haven to Pembroke, and they created some astonishing things the world had never seen. Each of these inventors also possessed the traits of practicality and persistence, two habits often ascribed to Maine people. Sometimes with only their spare time to give, it was stick-to-itiveness that kept them working through trials and disappointments to make real what their imaginations had conceived. Very often, their ideas were driven by the need to find better ways to do old and necessary tasks. That was what inspired Joseph Peavey to make a woodsman's tool, Charles Emery to solder tin cans, and Leon Leonwood Bean to make boots.

A number of these inventors were not only prolific, but also remarkably versatile. Margaret Knight, best known for creating a better paper bag, registered dozens of her other inventions, including a mechanical numbering machine and a ready-built window and sash. Hiram Maxim, whose name is forever tied to the machine gun, is credited with some two hundred more peaceful creations, ranging from a fire sprinkling system to a curling iron; and among the 130 inventions created by earmuff king Chester Greenwood were a steel leaf-rake and a folding bed.

Beyond their notable achievements, it is intriguing to consider how some Maine inventors flirted with even greater fame. In his determination to find time to perfect the reaper, Obed Hussey sold his candle-making business to a young assistant who turned it into an empire; Maxim was only days behind Thomas Edison in seeking a patent for the incandescent light bulb; Charles Lamson might have beaten the Wright brothers in the quest for flight; and the restless tinkerer Leonard Atwood might have found immortality had he not sold his elevator designs to Elisha Otis.

Not every Maine invention was born of necessity, and there were other enduring creations that were simply pleasing. Many are now taken for granted, but fairness suggests that some Maine credit is due whenever we enjoy a chew of gum, eat a tasty doughnut, sip a Moxie, or use a wooden toothpick to dig out that last morsel of lobster when a meal is done.

—Earl H. Smith
Belgrade Lakes, Maine
March 2021

Some say Obed Hussey was defeated by his own mild-mannered Quaker benevolence, but business competitors claimed the patch-wearing man was a "one-eyed devil" and he fought them with a ruthless vengeance.

1

War of the Reapers

By the early 1800s, American farmers were desperate to find ways to harvest their crops. Years earlier, the cotton gin brought an agricultural revolution to the South. Now, in the North, the clearing of forests was turning subsistence farms into vast acres of crops, and in the West, new fields of grain stretched on for miles. Harvesting by hand was impossibly slow, and while many inventors tried to build a mechanical reaper, the first practical machine can be credited to a Maine man.

Born in 1790 to a Quaker family in Hallowell, Obed Hussey (1792-1860) registered the design for a horse-drawn grain reaper with the U.S. Patent Office on the last day of 1833. At forty-three, he had already experienced a life of adventure and discovery. As a boy, he left Maine for Nantucket and worked aboard whaling ships in the Pacific. Years later he wrote of being lost at sea after a whale capsized his chase boat. Having lost an eye in another shipboard accident, he forever after wore a patch.

Obed Hussey

In the early 1830s Hussey moved to Baltimore where he worked in a tool factory and, during the harvests, in the fields. His determination to build a reaper was later explained in a letter to a friend: "I never experienced half the fatigue in rowing after a whale in the Pacific Ocean . . . " he said, "as I experienced year after year . . . in the harvest field."

His eagerness to mechanize farms was already evident, as he had previously patented machines that husked and ground corn, crushed sugar cane, and made artificial ice.

Hussey knew from the outset that his home state was not the place to test his reaper. Never mind that much of the land was hilly and difficult to plow, the real curse was the rocks, which often broke the shafts and blades of his machines. Even the rich bottomlands near Maine's rivers were strewn with glacial rocks that were painstakingly cleared to make open fields and build the many miles of handsome walls that defined them. Some erratic stones were too large to be moved and had to be left alone and plowed around, and as any modern-day farmer will attest, despite annual culling, the confounding rocks seem to grow like potatoes, year after year.

William Procter

Obed had no sooner settled in hilly Maryland than he moved farther west, to Ohio, where in Cincinnati he found not only flatter land but also the financial backing for his idea and a manufacturer willing to make the parts.

While he labored every free hour to perfect his precious machine, he supported himself as a tallow chandler (seller of candles), using a candle-making machine of his own creation. The business was a success, and might have led to a continuing career, but instead Hussey gave the enterprise over to his assistant in order to spend more time perfecting his field machine. The assistant was a newly arrived British immigrant named William Procter (1801-1884), who soon joined with his brother-in-law, James Gamble (1803-1891), and in 1837 formed what became Procter & Gamble, a consumer goods giant with more than $70 billion in annual revenue.

The Rival

In the spring of 1834, after Hussey's machine was roundly praised in the nationally circulated *Mechanic's Magazine*, the editors promptly received an irate letter from a Virginia blacksmith. Twenty-five-year-old Cyrus Hall McCormick (1809-1884) had been unaware of Hussey's invention and his letter challenged the validity of Hussey's patent. He insisted that he and his father Robert had actually developed the first grain reaper in 1831.

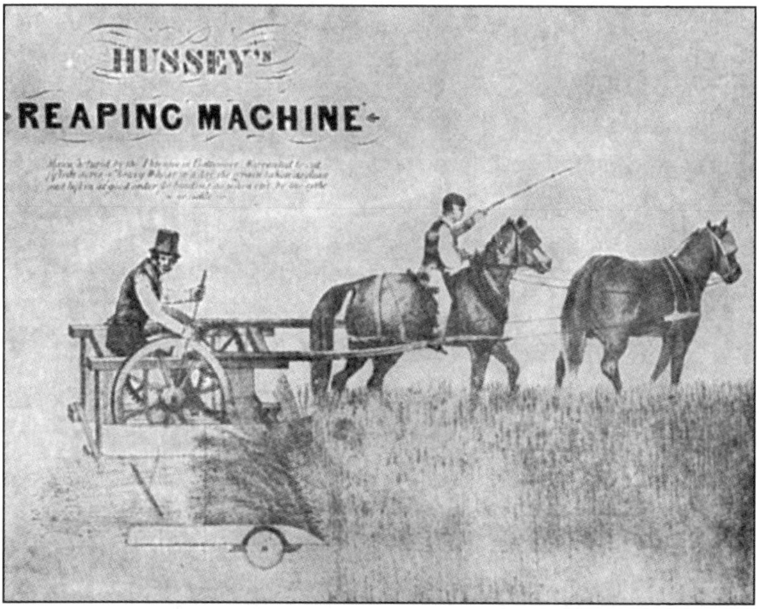

An illustration touting Hussey's reaping machine.

Hussey and the McCormicks were certainly the first to make practical reapers by implementing reciprocating (back-and-forth) cutting blades, but other inventors had long been scrambling to find ways to speed the harvesting of grain. Earlier machines, using scissors, were made in England and Scotland in the eighteenth century, and in 1826, a Scotch minister, Patrick Bell, made a reaping device, pushed by horses, but it was never widely adopted. In the United States, no fewer than seventeen patents associated with the reaper were issued before Hussey's, but most were models and none were ready for field use.

Upon learning of Hussey's claim, McCormick promptly paid the $30 registration fee, and President Andrew Jackson signed his patent on June 24, 1834. The comparability of the two designs instantly prompted what became known as the "War of the Reapers," a series of battles for commercial supremacy that were fought in the grain fields and in the courts for the next quarter-century.

With his seven-month lead, Hussey's machine was first demonstrated in 1835 in a barley field in Hamilton County, Ohio, where it efficiently cut and piled the stalks for bundling. Encouraged by his success, in 1836 Hussey moved his venture back to Baltimore, where he sold the first reaper in the United States before moving on to capture the markets in Ohio, Illinois, Maryland, New York, and Pennsylvania. However, McCormick, who had yet to sell a machine outside of Virginia, was unrelenting in his efforts to beat Hussey.

Although each man was a creative genius, there was otherwise little to compare the two. Hussey, a devout Quaker, was a refined man and a philosopher. He published two volumes of poetry and four journals of his life as a whaler. His satisfaction came from making new things, and he stubbornly refused to incorporate any parts into his reaper that he did not design. McCormick was more pragmatic. He cared less about invention than he did about sales, and became a marketing guru who was ahead of his time. By the early 1840s, McCormick was demonstrating his reapers at agricultural fairs throughout the country and offering free trials, a fixed price, installment payments, and money-back guarantees. Unlike Hussey, McCormick had no qualms about incorporating parts designed by others, and soon added improvements, most notably a divider that separated the cut grain and cast it to the rear of the machine for raking and binding.

War of the Reapers Begins

The first War of the Reapers was set for June 30, 1843, near Richmond, Virginia. Both men arrived three days early, and Hussey challenged McCormick to a pre-trial. McCormick went first. Before Hussey got his turn, it began to rain, and his reaper was unable to cut the soaked wheat. McCormick was declared the victor. The rain also washed away a nearby bridge, and when the day of the main competition arrived, Hussey could not get his large machine into the field and had to compete with a smaller one. He lost again, but following his protest, it was agreed to hold yet another battle the following week. This time, Hussey brought his large machine, but the wheat, tangled by the rain, broke both his sickle and rake. McCormick fared no better with his machine and the contest was declared a tie. In the aftermath despite the tie, Virginia farmers bought only two of Hussey's machines, while McCormick sold twenty-nine.

Each man's initial patent was set to expire in fourteen years. Hussey missed the 1847 deadline for reapplying, but that same year he received a separate patent for the invention of "guard fingers," installed behind the reaper's cutting slots to prevent the escape of particles that fouled the blades. The improvement was significant, and Hussey, in his objection to McCormick's application for a patent extension, claimed McCormick had copied it. McCormick's objection was denied, and the only appeal left for either man was through Congress or the courts. Congress refused to act and the courts were slow. The war moved overseas.

When the competitors demonstrated at the first world fair, the Great Exhibition in London in 1851, McCormick's reaper took the prize, but Hussey won a second contest in September, after which Prince Albert gave him the blessing of the crown by ordering two of his machines. The war then moved to France, where McCormick's invention prevailed at the 1855 International Exposition and the 1856 Agricultural Congress. Although McCormick was the winner in an 1857 field trial near Budapest, Hussey led the market that year by producing a record 521 machines. It was the last time he would hold the edge.

McCormick had moved his factory to Chicago in the 1840s, and a decade later was well poised at the railhead of the American West to win the reaper war. Hussey soon saw the handwriting. The Financial Panic of 1857 was limiting investment capital for expansion, and the nation itself was on the brink of civil war. In 1858, Hussey surrendered for $200,000, and he sold his patent to McCormick and a syndicate of attorneys, who set about to manufacture an improved mowing machine.

The following year, the court acknowledged McCormick's theft of the guard fingers and awarded Hussey $80,000 and patent rights thereafter, but it was too late for Hussey.

Some say Hussey was defeated by his own mild-mannered Quaker benevolence, but business competitors claimed the patch-wearing man was a "one-eyed devil" and he fought them with a ruthless vengeance. Whatever the truth, there is no doubt that his final act came from kindness. In August 1860, he and his wife and infant daughter set out by rail from Boston for a trip home to Maine. At a stop in Exeter, New Hampshire, he left the train to fetch water for the child of a fellow passenger. When he returned, the train had begun to move, and in his haste, Hussey fell under the cars and was killed.

Hiram and John Pitts

Throughout the nineteenth century, other Maine men were making headway in the promising business of grain harvesting. In 1834, the year after Hussey registered his reaper, Hiram A. Pitts (1799-1860) of Winthrop, son of a local blacksmith, began patenting a series of inventions for the thresher, a machine that followed the reapers and separated the grain. His twin brother, John (1799-1859), began making threshers as well, and established what became the Buffalo Pitts Company in New York in 1837. His invention of an attachment that counted bagged bushels won a gold medal at the Paris Exhibition of 1855. The brothers later licensed their inventions to other manufacturers, including the J.I. Case Company.

William Deering

William Deering

William Deering of South Paris (1826-1913) in the late 1800s took harvesting to a new level. Deering began his working life in textiles, and as co-founder of Deering, Milliken & Co. in Portland, made cloth and uniforms for the Union Army. He later moved to Illinois, where in 1878 he partnered with E.H. Gammon to manufacture grain harvesters. Deering's own inventions included the first self-propelled (steam) farm machine, which received a gold medal in the Paris Exhibition of 1900. Two years later, the Deering Harvester Company merged with the company founded by Hussey's old foe McCormick to create International Harvester. The deal was financed by J.P. Morgan. International Harvester would emerge as an iconic

A salesman demonstrates a McCormick-Deering tractor, produced by International Harvester, in Ukraine in 1931.

American company that produced agricultural equipment, automobiles, commerical trucks, lawn and garden equipment, and more.

History fails to record the name of the brave diver, but one can only imagine what he must have been thinking when he looked through the foggy glass of his helmet while being lowered into the cold river in a clumsy rubber suit with iron weights in his boots.

2

Submarinus

With hundreds of miles of coastline, five major rivers, and more than a million acres of inland lakes, the new state of Maine was ripe for inventors who sought to find new ways to deal with the water. In the early nineteenth century, most prominent among them was Readfield-born Leonard Norcross (1798-1865), a designer wizard immortalized by the creation of the world's first practical diving suit.

Norcross was only three years old when his father died in 1801. His impoverished mother apprenticed him to a nearby farmer who treated the youngster as a serf. The family rescued him, and he was sent to live with an aunt in Greene, who insisted he attend public school while training as a millwright. He was working at a Livermore sawmill in 1823 when, at twenty-five, he patented his first invention, a machine for shaping and finishing cut stones. A year later his second patent was issued for a machine that made wrought-iron nails.

In the early 1830s, Norcross moved to the Oxford County mill town of Dixfield, where the Androscoggin River meets the Webb

River. It was here that he became interested in making a suit that would allow divers to recover lost objects. In mill towns everywhere, river and sea bottoms were littered with all manner of equipment lost to angry waters or by clumsy hands. Whether made of iron or wood, when these things fell out of the reach of oxygen, they neither rusted nor rotted and were well preserved. The problem was, there was no way to retrieve them.

In what must have been a workshop cluttered by the detritus of other inventions, Norcross set about to stitch together a suit of cloth, backed by elastic India rubber, to which was affixed a metal helmet with a watertight seal. His plan called for a set of bellows in a boat above to feed air through a rubber hose and inflate the entire suit. Exhaust air was discharged through the helmet, and cast-iron shot was placed in the boots to overcome buoyancy and keep the diver upright.

The suit was first tested in the Webb River in the early spring of 1834. Lowered and retrieved in a basket by a hand-turned crank, the diver was able to move freely about on the bottom, and the experiment was deemed a success. History fails to record the name of the brave diver, but one can only imagine what he must have been thinking when he looked through the foggy glass of his helmet while being lowered into the cold river in a clumsy rubber suit with iron weights in his boots.

Although other patents had been issued for underwater suits (two in the previous year), the Norcross invention was the first to utilize rubber, and his patent, issued on June 14 of that year, stands as the earliest and closest prototype to the modern gear for scuba diving.

Agricultural Inventions

Heartened by his success with the diving suit, Norcross moved on to create two more inventions within a year. On June 15, 1835, he received patents for both an accelerated spinner for hand-woven wool and a mechanical device designed to extract stumps and rocks.

Leonard Norcross.

Soon after, he joined the growing crowd of agricultural inventors, and in 1837 created an improved threshing and separating machine. When his application was beaten to the patent office by fellow Mainer Hiram Pitts, Norcross and his supporters cried foul, but to no avail.

Norcross taught school to supplement his income, and as his fame grew, he was often sought as a lecturer on astronomy, and sometimes worked as a part-time preacher, readily using the pulpit to rail against the evils of alcohol. He died in Dixfield in 1865, and although he left behind an eclectic array of modern inventions, any doubt about the one that pleased him most was dispelled when he named his son Submarinus.

Vulcanized Rubber

As a practical matter, the Norcross suits had short lives. Although Americans had been wearing rubber-protected clothing and boots since

An illustration filed with Leonard Norcross' patent application for the first rubber diving suit.

the early 1830s, the crude rubber-backed cloth cracked in the winter cold and melted in the summer heat. Decayed boots were often buried in back yards to avoid the stink. It wasn't until 1839 that a Philadelphia inventor named Charles Goodyear (1800-1860) mixed rubber with sulfur on a hot stove and created a substance with elasticity, strength, and stability—vulcanized rubber. However, Goodyear died in poverty in 1860. In 1898, Frank Seiberling began making rubber bicycle and carriage tires in Akron, Ohio, and named his tiny company of thirteen employees in honor of Goodyear, the man who saved the industry.

Though it all, Robert Lewis was always thinking of ways to make work better and easier, the hallmark of a good inventor.

3

A Better Way

Robert "Bob" Benjamin Lewis (1802-1858), a flamboyant man of mixed African and Native American descent born in 1802, became one of the first Black men to successfully obtain United States patents for his inventions. He was also a noted author and successful salesman, earning a comfortable income selling hair oil to enthralled crowds in Southern Maine.

Lewis was born in what is now the city of Gardiner, and appears to have spent his youth in the Gardiner area, including nearby Augusta. As an older adult, he lived in Bath. He attended some school as a youth and received more education while preparing to become a missionary (although there is no evidence he did so) as an adult. That preparation gave him a strong knowledge of the Bible and classical history. As a young man he sailed on ships that worked the Atlantic rim and traded as far as the Caribbean. He also worked on local farms and frequently as a whitewasher. Through it all, he was always thinking of ways to make work better and easier, the hallmark of a great inventor.

Robert Benjamin Lewis.
Courtesy of the American Antiquarian Society, Worcester, Massachusetts.

By his early thirties, however, Lewis was earning a comfortable living selling his "inimitable hair oil" which was guaranteed to make hair grow and stay in good condition. To prove that his hair oil worked, he showed off his long black curls that stretched down below his coat collar and were heavily greased with his oil. According to the book, *Maine's Visible Black History*, Lewis "held the enthralled crowd spellbound with the description the potency of the product." The crowds in Maine bought bottle after bottle.

While his special hair oil provided him with income, his more lasted fame came from inventing and writing.

Oakum

In the early nineteenth century sailing ships, the rigging was made from hemp coated with pine tar and varnish and eventually needed to be replaced. The discarded rope was untwisted to single strands, infused with tar and loosely retwisted to create Oakum, which was then used to caulk the seams of wooden ships to make them water tight. The process of untwisting the hemp was a slow, tedious job usually given to the people low in the pecking order, or even sometimes to inmates, people who lived in poorhouses, or sailors who were being punished. Most likely, Lewis was also tasked at one point with the job. His solution was to invent a better way. On April 11, 1836, he filed a patent application for Robert Benjamin Lewis' Oakum and Hair Mill, which he described as a "new and useful machine" that would mechanize the work of the picker by using teeth on a cylinder that was turned by a crank to separate the strands (or hair) of rope. The invention became an important product at shipyards for years to come.

But Lewis was not done. He also received a patent for a machine that cleaned and restored the feathers used to adorn high-fashion clothing and a patent that improved the brushes used for whitewashing, a popular type of painting during the era.

As an author and historian, Lewis wrote *Light and Truth*, which was published in Portland in 1836. The 176-page book is perhaps the first history book about African Americans and Native Americans and is considered a foundational text for Black Nationalism. It included short sketches of prominent black leaders and denounced notions of white superiority. An expanded 400-page edition of the book was published in 1843 and the book would see two more printings.

In 1858, Lewis sailed as a cook and steward on a ship bound from Bath to Port-au-Prince in Haiti. When he arrived, he fell ill and died. He is buried in Haiti.

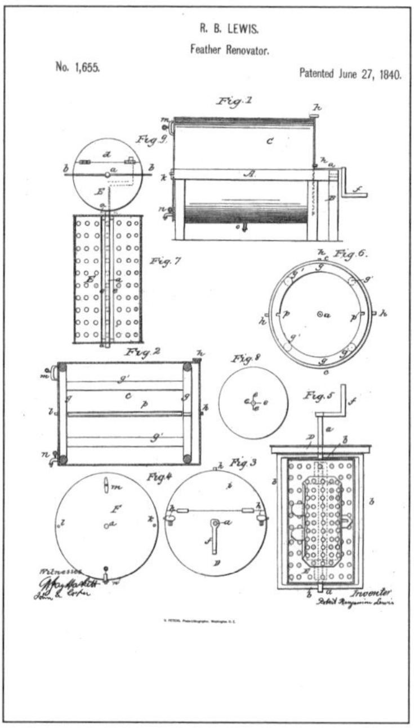

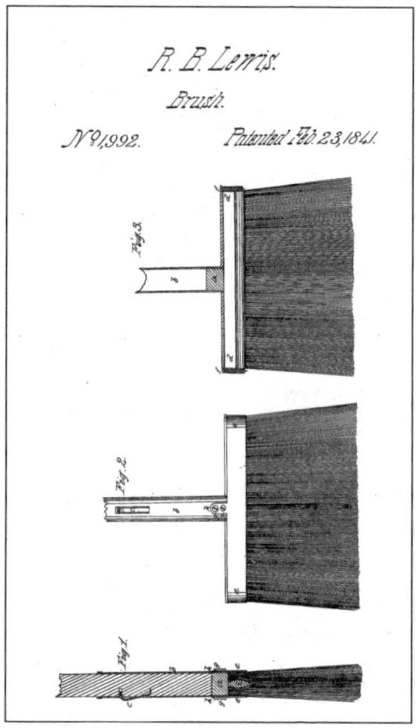

Illustrations from Robert Benjamin Lewis' patent applications: (left): Feather Renovator; Brush; (right): Hair Picker, 1841.

R. B. LEWIS.

9771X

Hair Picker.

Patented June 25, 1836.

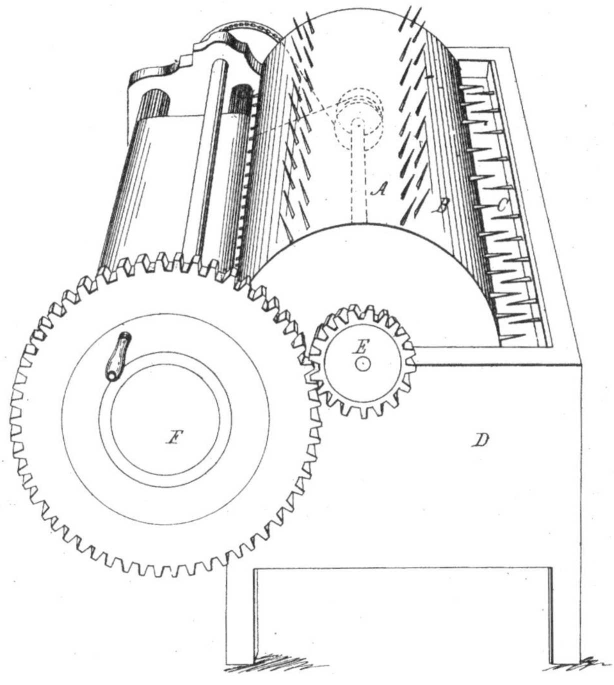

THE NORRIS PETERS CO., PHOTO-LITHO., WASHINGTON, D. C.

Those who knew Joseph Peavey
were not surprised by his creativity.
Although a blacksmith by trade, he
was a tinkerer at heart.

4

Praise the Peavey

Of the 2,686 inventions licensed by the U.S. Patent Office in 1858, most were creations meant to improve manufacturing schemes in the hundreds of new factories and mills. Maine people could claim a fair share of credit for the wonderful new gadgets, but perhaps the state's most remarkable contribution was the enhancement of a simple tool used in the harvesting of timber, the oldest Maine industry of all.

The inspiration for the new tool came during the spring freshet of 1857, when fifty-eight-year-old blacksmith Joseph Peavey (1799-1873) watched from a bridge as river drivers struggled to free a logjam in the water below. The logs were headed down the Upper Stillwater branch to the Penobscot River, where they would be carried on the "River of Logs" to Bangor and sawn into lumber at one of more than a hundred water-powered mills. In that year, some one-hundred and-fifty million board feet of lumber were shipped on more than three thousand ships from Bangor, the Queen City of the East and the largest lumber port in the world.

Joseph Peavey
Courtesy of Peavey Manufacturing Company.

The rivermen used cant dogs to pry at the jumbled logs. The wooden handled levers with steel hooks (dogs) at the end had been used for ages to turn and lift logs on the drives, load the sleds, and position canted logs (square on one or more sides) at the mills. Here in the furious water, Peavey could see that the swiveling "dogs" at the end of the poles made it difficult to gain purchase on the treacherous logs, many weighing a thousand pounds or more and driven by the force of an enormous current.

Peavey perceived a better way, and when he returned to his smithy, directed his son, Daniel, to refashion the tool's hook by giving it a spike at the tip and securing the dog to allow it to move up and down, but not sideways. His creation was first tested by a river driver, William Hale, who declared it a marvel, and with its eponymous new name, the Peavey soon became the favored tool of woodsmen everywhere.

Those who knew Joseph Peavey were not surprised by his creativity. Although a blacksmith by trade, he was a tinkerer at heart. Before creating the improved logging hook, he invented a clapboard water wheel, and a "Peavey hoist" for pulling stumps and lifting gates at dams. His other patents included a hay press, and a vice that utilized a wooden screw. Having undoubtedly had more than one mishap at the drawing board, he also patented an inkwell that could not turn over and spill.

Bangor Edge Tool Company

The small Peavey shop in Stillwater was quickly swamped with orders for the new tool, and Joseph and his sons, Daniel and Hiram, moved the operation to Orono for a time, and then upstream to Old Town, where production continued until Joseph's death at age seventy-

The Bangor Edge Tool Company was founded in 1873 by brothers James H. and Charles A. Peavey in Bangor. James and Charles were grandsons of Joseph Peavey.
Courtesy of Peavey Manufacturing Company.

four in 1873. Later that year, Joseph's grandsons, Charles and James, opened a shop on Exchange Street in Bangor where the Bangor Edge-Tool Company sold the peavey along with a variety of other woodsmen's tools. (About that time, James made further improvements to the peavey by combining the tool's socket and the spike.)

At the end of World War II, the Maine Woods Tool Company (MWTC) made the peavey and other woods tools in Oakland and Madison, and in 1954 the two companies merged. Following two devastating fires, the Peavey Company moved to an abandoned chicken

Paul Bunyan statue with a peavey on Main Street, Bangor.

house in Eddington (Route 9), where it continues to make the simple, useful tool still sought by woodsmen around the world.

The importance of peaveys to Maine's woodland industry was immortalized in 1925 by Brewer-born American artist Charles (Carl) Eugene Tefft, best known for his sculpture of Maine's Hannibal Hamlin in Statuary Hall at the U.S. Capitol, his bronze and granite sculpture titled "The Last Drive" stands near the Bangor Public Library on Harlow Street. A gift to the city from the family of Bangor lumber baron Luther Pierce, the work memorializes legendary river driver Pat Connors and a young logger, David Preble. The sculpture prominently displays the men separately holding an axe, a cant dog, and a peavey.

Circular Saws

While the peavey vastly improved the efficiency of moving timber, the circular saw doubled the speed of cutting it into lumber. There is dispute about who invented the circular saw, but it is likely that the first one seen in Maine was around 1845 in Waterville, when William Kendall, son and namesake of the Revolutionary War general whose name was given to Kendall's Mills (later Fairfield), riveted teeth to a six-foot disc of boilerplate and set it up to slice logs at his sawmill at Ticonic Dam on the Kennebec River in Waterville. It could do the work of four reciprocating saws that wasted half their motion, and people came from miles, willing to pay a small fee to the enterprising Kendall to watch it work.

At the same time, another Maine man, James Emerson, was taking saws a step further. A native of Norridgewock, he ran a factory making woodworking machinery in Lewiston before moving to California in 1852 where he improved both the band and circular saws by adding the feature of removable teeth that could be sharpened or replaced with relative ease. He returned east to Trenton, New Jersey, in time to make swords and sabers for the Union Army, and then became superintendent of the tool-making American Saw Company.

Helen Knight's obituary hailed her as "a woman Edison," and long after she died she stood as a role model for women who desired careers in the male-dominated fields of mechanics and invention.

5

A Better Paper Bag

In the years following the Civil War, Americans began searching for ways to expand the multiple technologies that were born in the strife and to discover new ways to improve their daily lives, at home and in the workplace. As swords became plowshares, a second Industrial Revolution brought to the U.S. Patent Office a flood of new creations from the best minds in every walk of life. However, very few of the patent applications were from women.

The lopsided representation of the sexes was not a reflection of women's ingenuity, for women had been creating things all along, and throughout the Civil War, women had conceived a myriad of inventions, simply to survive. Neither was the disparity the fault of the Patent Act of 1790, which clearly stated "any person" could petition for protection. Instead, the greatest barriers were the laws of the separate states that made girls and women property of their fathers and husbands. Without their own property, women had little means to pay for patents, and moreover, it made little sense to protect the things they could not own.

Men had already claimed more than five thousand patents by 1809 when Mary Kies (1752-1837) of Killingly, Connecticut, became the first female patent holder for her technique of weaving straw and silk together to make fine hats. Newly elected President James Madison signed her certificate; his wife, Dolley, bought one of her hats and then sent a note of congratulations. Kies' fortunes soared for a time, but eventually the hats went out of fashion and the first woman to hold a United States Patent ultimately died penniless in 1837. She was buried in a pauper's grave.

Kies' historic patent materials, along with those of some ten thousand others, were destroyed in an 1836 fire that swept Blodgett's Hotel, a Washington, D.C., building (never a hotel) shared by the Patent Office and the US Post Office. Maine US Senator John Ruggles (1789-1874) of Thomaston was chair of the Committee on Patents and Patent Office, and he led the effort to rebuild and reorganize the agency as the U.S. Patent and Trademark Office and place it under the aegis of the Department of Commerce. Although design models of many of the early inventions were lost in the fire, most of the patents were recertified and then serially numbered with the prefix X. The first (X1) patent issued by the reorganized agency was to Ruggles himself, for his design of a train wheel made to withstand the winter weather.

The patent playing field for woman did not begin to level until 1839, when Mississippi became the first state to depart from common Law by adopting the Married Women's Property Act that allowed married women to own property. Slowly, other states followed with their own versions of the act. In 1844, Maine granted women the right to hold trade licenses and own property, but not control it. In 1857, that law was amended to allow women to not only to control their own earnings, but also to make contracts. By the end of the Civil War, twenty-nine states had taken similar steps forward.

Margaret E. Knight

The new legislation helped, but the struggle for equality for female inventors was far from over. Social norms still gave men wide advantages, and most women lacked the financial means or the scientific training to make their way into the elite world of patent holders. While utility and design patents were issued at an ever-increasing rate through the 1870s, all but a handful were awarded to men. One of the women who managed to break through in the male-dominated world was Margaret E. Knight (1838-1914), who became arguably the most famous woman inventor of the nineteeth century and is said to have claimed more than 80 patents.

Born in 1838, Knight seemed destined to become an inventor from the get-go, and while she had little formal education, she taught herself the skills she needed to build the things she dreamed of. Growing up in her hometown, she was often derided as a tomboy, but the "slur" did not keep the local boys from admiring her cleverness and begging her to build them kites and sleds.

Knight told the *Boston Sunday Post* in 1912 that her first invention occurred when she was just twelve-years-old. Her family was living in Manchester, New Hampshire, and she was in and out of the local cotton mill, sometimes bringing lunch to her brothers.

One day, she saw a woman badly wounded when a dislodged spindle flew off a loom. Horrified, Knight set to work designing a safety device, what she called "a simple contrivance," that could stop a loom when a weaving thread was broken. Although it was never

First Woman Granted an American Patent
Is Still at Work on Ingenious Inventions

MISS MARGARET E. KNIGHT AT WORK IN HER EXPERIMENTING ROOM.

Boston Sunday Post, March 31, 1912.

patented, the safety mechanism became widely adopted in textile mills everywhere.

"Thus you see that it was early in life that I showed an inclination to tinker and experiment with machinery," she said. "There was nothing in that first invention for me from a monetary viewpoint, but it increased my desire for further study of machinery."

The Flat-bottomed Paper Bag

Knight was never married, and while working to support herself, she continued to build and tinker. In 1870, at the age of thirty-two,

Some Patents of Margaret Knight

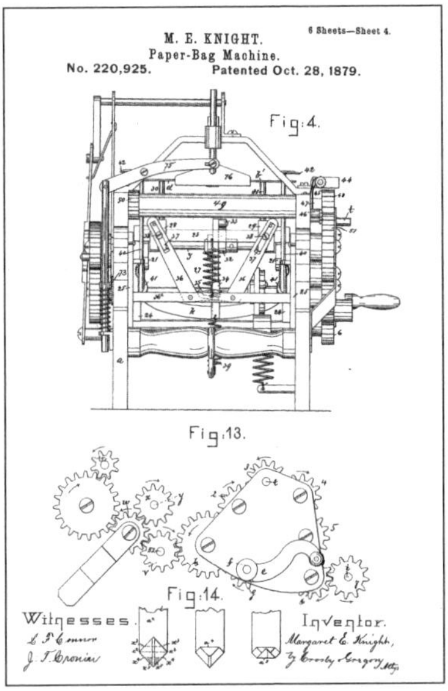

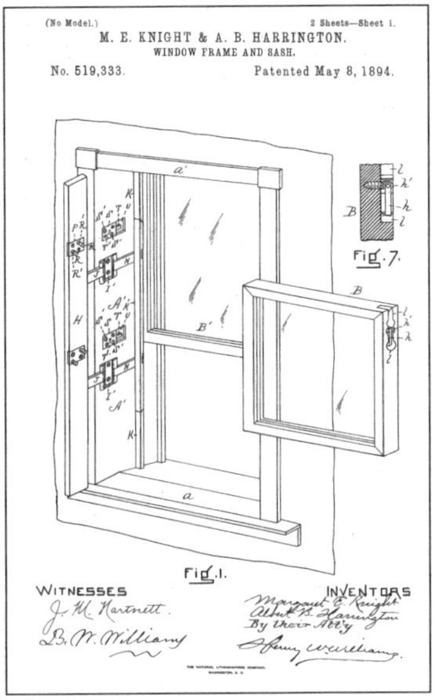

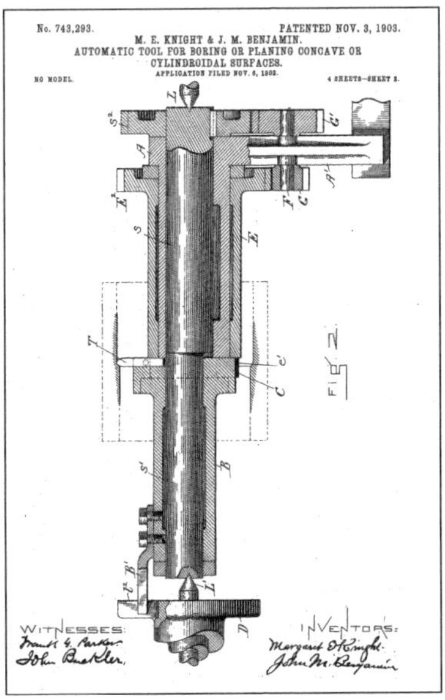

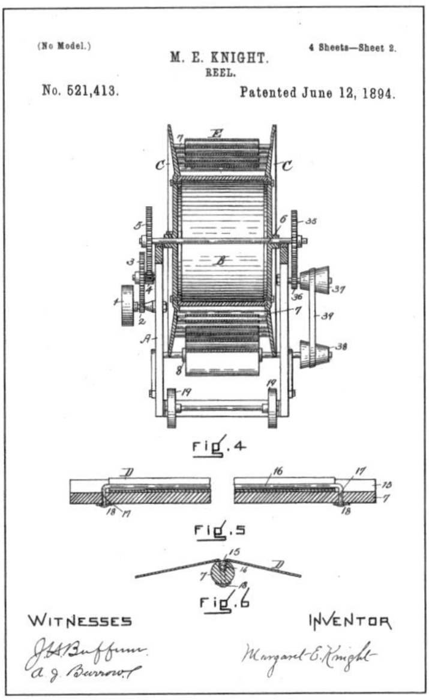

Margaret Knight registered numerous patents during her lifetime. They included (clockwise from top left): Paper-bag machine, 1879; Window frame and sash, 1894; Reel, 1894; Tool for boring or planing concave or cylindroidal surfaces, 1903.

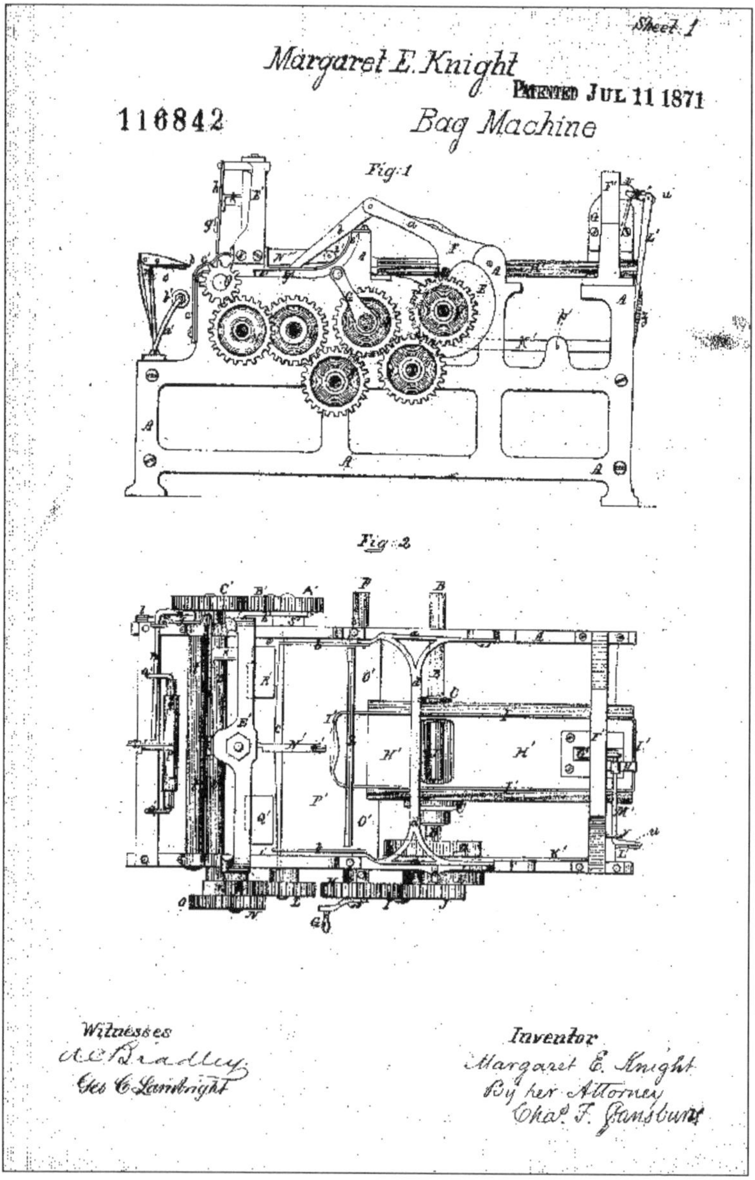

Illustration of a new machine for making flat-bottom paper bags, 1871.

she designed a machine that made square-bottomed paper bags, a welcomed and revolutionary creation that made her famous.

Although hand-carried bags had been around for centuries, the universal means of lugging produce from the markets was the basket, most were made with handles and shaped from tree bark or woven from local reeds and grasses. Every farm and city home had baskets of all kinds and for every purpose, some baskets were even woven tightly enough to carry water.

Francis Wolle (1817-1893), a Moravian priest from Pennsylvania, made the first paper bag in 1852. Flat and straight-sided, the bag was accepted, but had its failings—It was flimsy, tore easily, and was not universally adaptable. Consider, for example, putting apples in an envelope.

In the late 1860s, Knight was working at the Columbia Bag Company in Springfield, a maker of Wolle's slender bags. The company was looking for ways to make a flat-bottomed sack, and agreed to Knight's request that she be allowed to help them design a machine to make one. The design process was slow work, and when her boss complained that she was spending too much time on the project, she offered the company the rights to whatever she created, if she was allowed to continue.

The first machine she devised was made of wood, and it successfully cut, folded, and pasted paper bags with flat bottoms. An iron prototype of the machine was then made in Boston, the first step in meeting the requirements for a patent application.

As sometimes happens when a woman begins to climb the ladder, a man tried to get in the way. Charles Annan had apparently visited the bag factory and got a look at Knight's wooden model. When she made her patent application, she discovered Annan had already filed for a suspiciously similar machine. Although she had little money to spare, she took him to court just the same. Knight's case was built on her working notes, diary entries, samples, and the affidavits of factory witnesses. Annan's defense was that no woman—certainly not an uneducated one—could build such a complicated machine. The court found for Knight.

Having persevered in court, she then took another further step toward justice when she objected to the patent office tradition of identifying patent holders solely by first initials and last names. She wanted it clear to all that the flat bag inventor was a woman and she insisted that her name be listed as Margaret E. Knight, not M. E. Knight. It was not the last time she fought a patent infringement in court and in 1912, she claimed that she never lost a case.

The history of the paper bag is long and wrinkled. Following Knight's invention of a bag that could stand up, it gradually became the popular mode of carrying goods from markets. For the most part, baskets were retired to walls for decoration. The trend continued for almost a century, until the 1960s when grocers began handing out new plastic bags. The use of plastic bags was a move aimed at saving money and trees; both capitalists and environmentalists cheered. Alas, by the 1980s, the world began to realize that polyethylene plastic bags were helping kill the planet faster than deforestation, and after a decade of debate, grocers began urging customers to buy their own reuseable bags. Even then, the saga wasn't over. When the COVID-19 pandemic struck in 2020, stores reverted to issuing both plastic sacks as well as Margaret Knight's paper bags. Grocers feared reusable bags might help spread the virus.

Over her lifetime, Knight was known not only for the wide-ranging variety of her inventions, but also for their sheer number. Her list of creations included a dress shield to protect material from perspiration, a clasp for robes, a ready-made window frame with sash, a numbering machine, a half-dozen mechanisms to improve the manufacture of shoes, and following the introduction of the rotary engine in 1902, several devices that advanced its development. In the later part of her career, she spent a lot of time working with motors.

Despite her success and some renown, she perhaps never found the wealth and fame she might have because she shunned publicity. She hated interviews and she hated to have her picture taken.

"I guess that I have never learned the value of publicity," she said in 1912. "For I never care to be interviewed. I like to produce results rather than talk about them."

When she died in 1914 she had claim to nearly one hundred inventions and she said in 1912 that she had at least 87 patents, twenty-seven of which were patented. Knight was inducted into the National Inventors Hall of Fame in 2006. A fully functional patent model of her original bag-making machine is located at the Smithsonian Museum in Washington D.C. Her obituary hailed her as "a woman Edison," and long after she died she stood as a role model for women who desired careers in the male-dominated fields of mechanics and invention.

After her father's death, Helen Blanchard's mother was forced to sell the family home and move to Boston. In order to help her family survive, Helen turned to her inventing skills in earnest.

6

Zigzag

By odd coincidence, another rare nineteenth century female inventor was a Maine contemporary of Margaret Knight, born only a few miles away, in Portland. Although the two were parallel in time and place, they came from opposite ends of the social hierarchy, and there is no evidence that the two ever met.

Unlike Knight, Helen Augusta Blanchard (1840-1922) came from wealth. Her father, Nathaniel, was a prominent shipping merchant, and his wife and six children prospered. Although her education did not include the sciences, she became fascinated by mechanical devices and dabbled at home with her own creations until tragedy struck. She was twenty-six when bank failings in England and France caused a global economic downturn. Her father lost his fortune in that Panic of 1866, and the tragedy left him in poor health and soon led to his death. Blanchard's mother was forced to sell their home in Portland and move to Boston, where in order to help her family survive, Helen turned to her inventing skills in earnest.

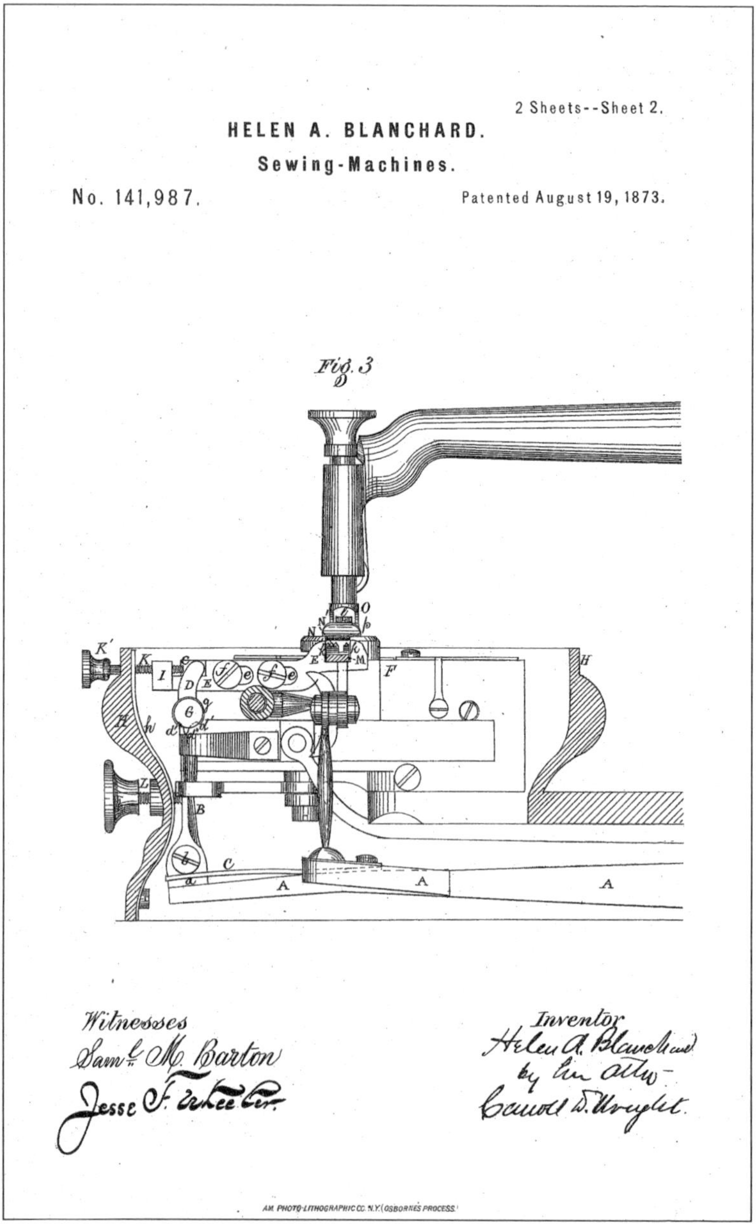

Patent illustration for Helen Blanchard's sewing machine, 1873.

Helen Augusta Blanchard

It was an era when the making of textiles was moving from homes to factories, and Blanchard seized the moment to invent improvements of machinery for the growing number of riverside textile mills in New England and beyond.

In 1873, she borrowed money to pay the fees for her first patent, a factory sewing machine that made sturdy buttonholes with zigzag stitches. Although she later made many sewing machine improvements, the zigzag was the one that is best remembered. Among her other contributions was a device for over-seaming that simultaneously stitched and trimmed knitted fabrics. Her early successes led her to Philadelphia, where in the early 1880s she established both an over-seaming and a hosiery machine company in her own name. A decade later, she moved to New York City where she continued her inventing successes.

Throughout her inventing life, she struggled with a slow and often unresponsive U.S. Patent Office. Her entry in an 1893 encyclopedia of prominent women included an explanation: "The ambition and energy that have marked her life were stimulated by the numberless annoyances and obstacles that always beset the pathway of a persevering inventor, in the shape of Patent Office delays, mercenary infringement of her rights and unscrupulous assaults on the products of her brain."

Blanchard undoubtedly agreed, but she persisted in the belief that sexual discrimination was at the root of most of the bureaucratic obstacles she faced.

In spite of the delays, during the course of forty-five years she was awarded twenty-eight patents, twenty-two related to factory sewing and sewing machines. Others included a hand-cranked pencil sharpener and an improved surgical needle.

Blanchard returned to Portland in 1901 and repurchased the family homestead. In her later years, she became an effective leader in philanthropic efforts focused on helping women who had been displaced by the machines she had helped to improve.

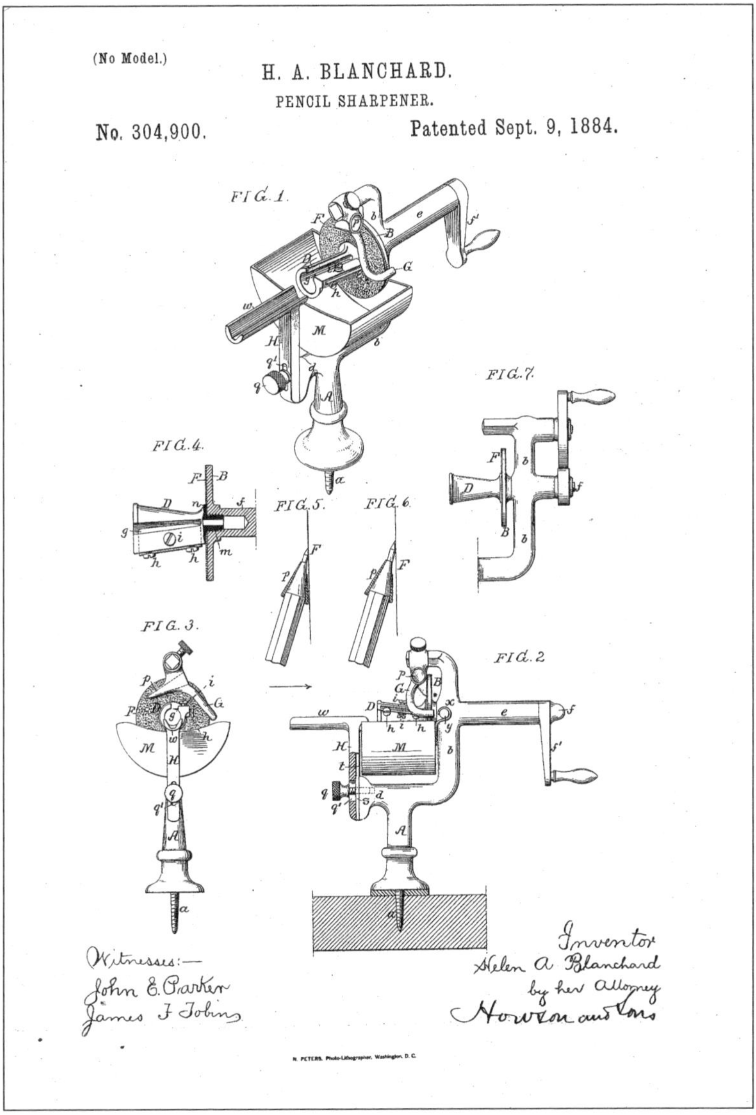

Patent illustration for Helen Blanchard's pencil sharpener application, 1884.

The story of who put the hole in the doughnut may be apocryphal, but don't tell that to the people of Clam Cove, where a bronze plaque immortalizes Hansen Gregory and his tasty creation.

7

Life's Simple Pleasures

In the mid-1800s, as the world moved out of the first Industrial Revolution and into the second, most of the thousands of newly patented inventions were mechanical devices and methods that improved the speed and efficiency of the country's expanding industries. Around the edges were a handful of new discoveries and creations that did little to directly advance the factories, but instead added to the personal pleasures of hearth and home.

Migration from rural farms to the city factories opened up a new world for those who were accustomed to working from sunrise to sunset. Shift work at the mills allowed for leisure time at home and even some disposable income, and before the days of the phonograph or radio, Americans were fascinated by parlor games.

Milton Bradley (1836-1911), born in the Kennebec County town of Vienna into a working class household in 1836, moved with his family to Lowell, Massachusetts, when he was eleven. He later attended Harvard's Lawrence Scientific School. In 1859, he

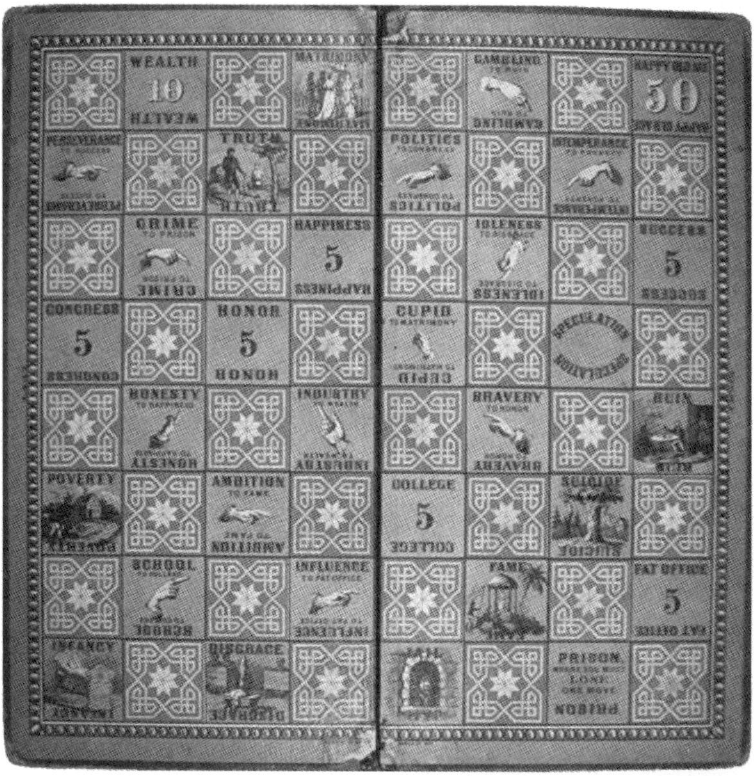

Above: The Checkered Game of Life. Right: Milton Bradley.

learned lithography in Providence, Rhode Island and then returned to Springfield, Massachusetts where he set up the first color lithography shop in the city. For the 1860 presidential election, he produced a very popular reproduction of a portrait of Abraham Lincoln. He also entered the board game industry in 1860 by creating the wildly popular Checkered Game of Life, later known as The Game of Life, which sold more than 45,000 copies in the first year. The Milton Bradley Company never looked back. Today, the company is owned by Hasbro and its games include Battleship, Candy Land, and Twister.

Hansen Crockett Gregory

The Doughnut Hole

In that same era, the game of life most certainly included the delights of other kinds.

The deep-fried pastry doughnut had been around for ages, having originated in the Middle East and eventually been rolled through Europe before being brought to America by the Dutch. As to the matter of the doughnut's hole, it seems to have been the creation of a Maine man, Hansen Crockett Gregory (1832-1921), the captain of a lime ship that sailed out of his home port in Clam Cove, now the Glen Cove section of Rockport.

There are multiple versions of just how Gregory came to invent the doughnut hole. In one, he was sitting in the kitchen and asked his mother why the center of the fried cakes she was making were so soggy. She didn't really know, so he proceeded to take a fork and punch out the middle of the cut dough and asked her to fry this one. In another story, told later in life by Gregory, he was on a schooner eating tough fried cakes when he decided to take the ship's tin pepper box and cut out the middle. Voila!

However, in the most exciting version, Gregory, who was indeed a Dutchman, perhaps lending credence to the story that he was eating a doughnut on June 22, 1847, when his ship was struck by a violent storm. In order to free both hands to manage the ship's wheel, he jammed his doughnut over one of the wheel spokes. When the calm seas returned, he reclaimed his impaled doughnut and discovered he had accidentally solved the long-held frustration of eating doughnuts with uncooked centers.

The storm story may be apocryphal and exact details about the doughnut hole scarce, but the people in Clam Cove have no doubts about who they think invented it: a bronze plaque in town immortalizes Gregory and his tasty creation.

Chewing Gum

About the time Gregory first pierced the doughnut, a second oral treat was coming out of the Maine woods. Lumberjacks had long been scraping spruce gum, essentially hardened sap, off spruce trees to chew. The problem was that whenever they sat quietly for a time and their mouth got dry, they risked having their teeth pulled from their sockets when they finally opened their jaws. The "gum" was tough and fairly tasteless.

Despite the rigors of chewing, woodsman John Bacon Curtis (1827-1897), who was born in Hampden, was addicted and wanted to improve the experience. Sometime around 1850, he struck upon

Above: Maine chewing gum packets. Right: John B. Curtis, circa 1890.

the idea which would make his habit a bit more pleasant. At the family home in Bangor, he filled an iron kettle with a mix of spruce gum, maple syrup, and a bit of sugar, and brought it all to a boil on the kitchen stove. The finished mixture was poured onto a large sheet of tin and sprinkled with cornstarch. When it cooled and hardened, he sliced it into tiny squares.

His State of Maine Pure Spruce Gum slowly became a hit as Curtis became the first to man to commercialize chewing gum. Eventually, Curtis added flavors to his gum, which only increased its popularity, and he moved the business from Bangor to Portland to better expand his production and markets. To keep up with demand, his company, Curtis & Son, developed a machine to mass-produce gum. Curtis would create, though never patent, many of the machines in his factory. Eventually, the company employed more than two hundred people and turned out eighteen hundred boxes of chewing gum a day. Some of the spruce gums sold were called "American Flag," "Yankee Spruce," "White Mountain," "Licorice Lulu," "Trunk Spruce," and "Sugar Cream." He, working in concert with his father, sold enough gum to build a dozen ships and buy the Forest City steamboat line in Portland.

Moxie

As if Maine had not already contributed mightily to satisfying the nation's taste buds, its ties to delectable diversions took yet another step after the Civil War, when patent medicines were all the rage. It began with a remedy called Moxie Nerve Food, and became a soft drink and a Maine icon.

Although the government did not patent the patent medicines, use of the term helped give credence to claims of the makers that their concoctions provided cures for every imaginable affliction ranging

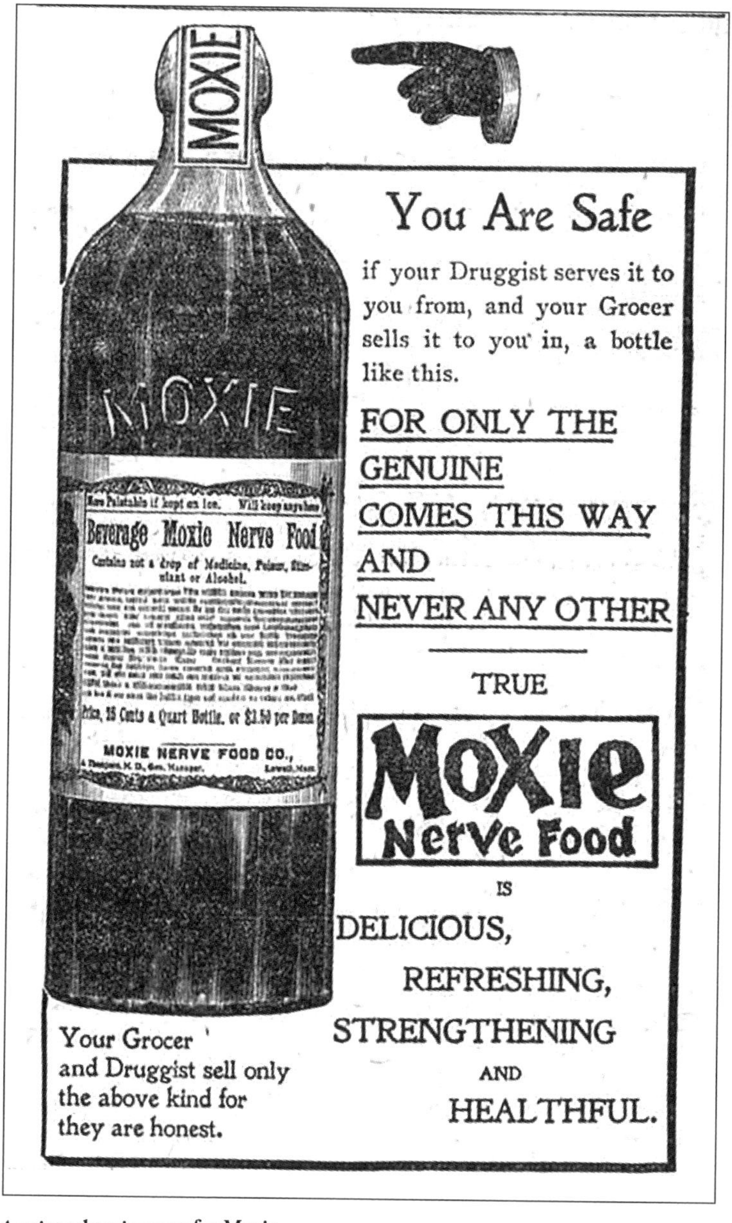

A print advertisement for Moxie.

from constipation to consumption, and from hay fever to headaches. Few actually worked. Many were laced with morphine, opium, or cocaine, and were downright dangerous.

Union native Augustin Thompson (1835-1903) served as a medical officer with the 28th Maine Volunteer Infantry during the war, and later practiced medicine in Lowell, Massachusetts. On a spring day in 1884, he was visited by a Union Army comrade, Lt. Clyde Ambrose Moxie, who brought with him a sample of an elixir he had procured in South America. He told Thompson that the natives assured him that the medicine gave the body exceptional power. Its essence was an extract from the bitter root of the herb called gentian.

Thompson wasted no time duplicating the recipe and marketing it as Moxie Nerve Food in honor of his friend. The label on the bottle guaranteed the nostrum would cure a range of afflictions including "loss of manhood, paralysis, and softening of the brain."

The touted nerve food, first dosed by the spoonful, caught on quickly, and in 1886 Thompson closed his practice and began selling his potion as a carbonated drink. First called Beverage Moxie Nerve Food, it quickly became simply Moxie. It was among the first soft drinks mass-produced in the United States.

The soda's unique flavor, sweet but with a bitter aftertaste, attracted thousands of customers, and by the turn of the century Moxie was outselling all other soft beverages in New England, including the burgeoning Coca-Cola. Because of its robust taste, the name, with a small 'm', entered the vernacular language meaning courage, daring, or chutzpah.

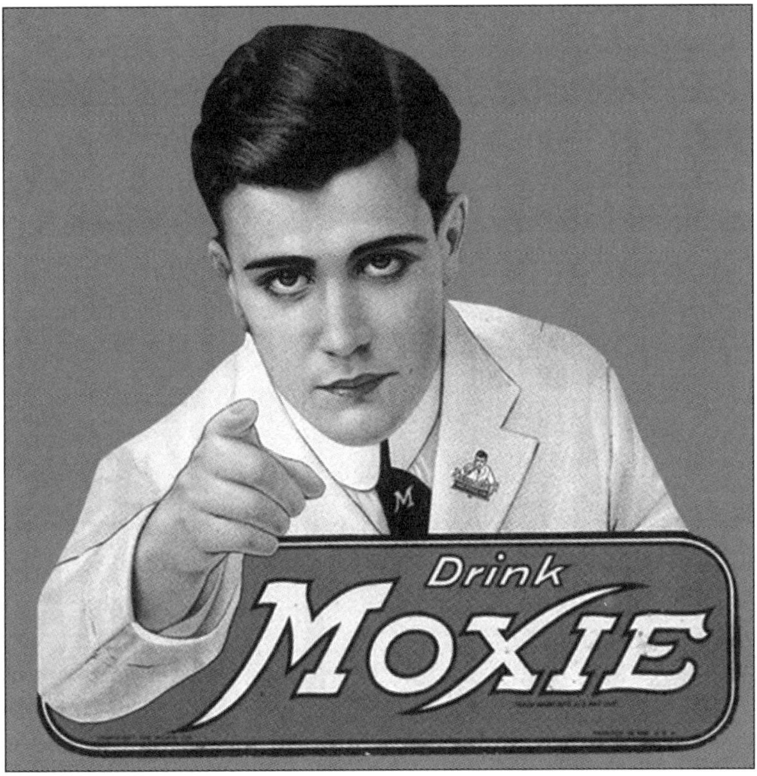

An old Moxie sign.

At one time, Boston Red Sox legend Ted Williams endorsed the product on radio and in print advertisements. President Calvin Coolidge also enjoyed Moxie. By the second half of the twentieth century sales had declined substantially from the drink's heyday.

Moxie became Maine's official soft drink in 2005, and in 2018 its old competitor, Coca-Cola, purchased the Moxie Beverage Company.

At the time of his death in 1937, the Smithsonian Institution ranked Chester Greenwood among America's 15 "most outstanding inventors."

8

Cover Your Ears

Chester Greenwood (1858-1937) had big ears. Not so big as to cause embarrassment, but far too big for a Farmington boy who liked to while away winter afternoons skating on frozen ponds. By 1874, at age sixteen, he had flirted with frostbite one too many times, and set about to find a way to avoid the suffering.

His creation was startlingly simple. With the help of his grandmother, he sewed a patch of beaver fur to each end of a piece of spring wire and clamped the apparatus on his head. He may have drawn a few stares in the beginning, but any ridicule quickly turned to envy. Barely two winters passed before most everyone in town wanted a pair of his cozy ear mufflers.

The young inventor was barely nineteen in March 1877 when he applied for and received a U.S. Patent, not for earmuffs—earmuffs had been patented before—but for the v-shaped hinge he had devised that kept the mufflers tight to the ears. With license in hand, he set up a factory on Front Street in West Farmington where the mufflers,

Chester Greenwood (right), his wife Isabel (left), and family at his home on Hill Street, Farmington.
Courtesy of the Farmington Historical Society and the George Greenwood family.

stitched at home by local women, were assembled. By 1833, the factory was turning out nearly fifty thosand pairs a year.

The Champion Ear Protectors Company soon became Farmington's leading industry and the tiny Sandy River town (population 3,300 in 1880) was the undisputed earmuff capital of the world. Farmington earmuffs were issued to American troops fighting in France during World War I, and following the war annual production soared upward each year, to nearly a half-million by 1936.

Although hooded jackets and caps with attached earflaps eventually ate away at the earmuff market, memories of its heyday are kept alive by Chester Greenwood Day, an annual Farmington celebration held on the first Saturday of December. Those who still like to wear earmuffs, along with others who resurrect them from their closets, join a gala parade to invigorate the chilly but charming center of town.

However the celebration is not entirely about earmuffs, as Greenwood's first invention was only one of many. Over his lifetime the skilled mechanic created some 130 devices. Many were patented. At the time of his death in 1937, the Smithsonian Institution ranked him among America's 15 "most outstanding inventors." Among his inventions was the now-ubiquitous steel leaf-rake, a shock absorber akin to the one used in modern airplane landing gear, an improved spark plug, a decoy moose trap, a donut hook, and the forerunner of today's folding bed. Along the way, he established and ran a successful bicycle shop, and, at the end of the century, co-founded with C.A. Allen the Franklin Telephone & Telegraph Company that manufactured its own transmitters and receivers.

Greenwood's wife, Isabel Whittier, was famous in her own right. An early suffragist, she vigorously campaigned on behalf of women before they had the right to vote. In conservative Maine, her success came not so much from a moral awakening as from the fallout of the Industrial Revolution her husband helped bring about. Women were taking jobs in factories everywhere and often dominating the work forces in Maine textile mills. Isabel Greenwood was among the few who stood to defend their working rights.

His marketing scheme was ingenious.
He hired Boston college students to visit stationers and ask for toothpicks, which they knew were not in stock, and to take meals in Boston's finest restaurants and request toothpicks when they paid their tabs.

9

Toothpicks

Despite what you might hear in Franklin County, Charles Forster did not invent the toothpick. Archaeologists assure us cavemen were using them in the Stone Age, and Emperor Nero sported silver toothpicks at the Forum. Forster's claim to fame is that he held the first U.S. patent on a machine that churned them out, and even more to his credit, found a way to get folks to use them.

Forster was born in 1826 in Charlestown, Massachusetts, and, as a young man, worked for an uncle in the import-export trade, sailing to ports around the world. Later, describing his start in the toothpick business, he explained that on his first trip to Brazil he was struck by the handsome appearance of the natives' teeth and was told the fine quality was the result of the faithful use of toothpicks. On that rather thin reed, he determined to make a fortune.

At first, his toothpicks were made by hand, through a labor-intensive process that was slow and awkward, but in 1869 Forster turned to Norridgewock native Benjamin Franklin Sturtevant (1833-1890) for help.

Wilton Forster Toothpick Mill Factory, circa 1940's.

Sturtevant, apprenticed to a shoemaker at age fifteen, was in business for himself in 1856 when he invented a spiral lathe that made the wooden shoe pegs used to attach the soles to the upper leathers. His machine could rapidly slice long, thin strips of birch or poplar veneer that were then cut and finished to make precise and uniform pegs.

In 1860, when Sturtevant opened his own shoe peg company on Sudbury Street in Boston, the nation was on the brink of war and his business struggled. Nine years later, with the war over, the fledgling company was still struggling, and Sturtevant was only too eager to sell Forster patent rights to allow him to adjust the machine to make the thin strips needed for creating toothpicks.

Forster's revised lathe easily sliced long, eighth-inch thick strips of white birch (or sometimes poplar) before an adjacent cutter uniformly chopped them into toothpicks. Once dried, the picks were smoothed in

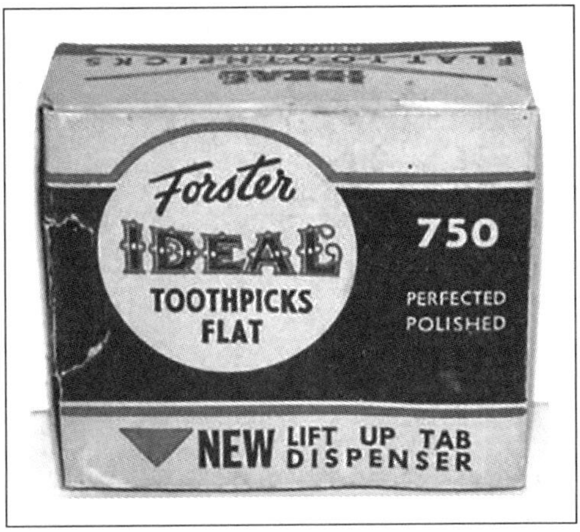

Vintage box of Forster Ideal toothpicks.

a heated drum with talcum powder, and then stacked in one direction before being finally packed in tiny boxes.

To be near a source of necessary hardwood, in 1887 Forster set up a factory in an old starch mill on Valley Brook in Strong and began turning out toothpicks by the millions.

It wasn't until he was swimming in toothpicks that he realized he had overlooked the basic economic theory of supply and demand. By then, he had a bazillion toothpicks and no one to buy them. He certainly didn't sell many where they were made. Mainers were far too frugal to buy anything they could easily make with a jackknife and a chunk of wood. In desperation, Forster set his sights on Boston, where there were fewer woodpiles and more free money.

His marketing scheme was ingenious. He hired Boston college students to visit stationers and ask for toothpicks, which they knew were not in stock, and to take meals in Boston's finest restaurants and request toothpicks when they paid their tabs. Confused merchants

Sturtevant Had Lots of Fans

Benjamin Sturtevant, 1888 by H. Badger.
Courtesy of Vincent Tocco Jr.

When Benjamin Franklin Sturtevant sold his lathe patent to Forster, he was already planning to get out of the shoe peg business. His veneer lathes created a great deal of dust, which were an annoyance and a hazard to the health of his employees. In 1864, he patented an exhauster, a powerful centrifugal fan that sucked the dust away. Quiet and efficient, the machine was a vast improvement over the earlier piston-driven exhaust systems, and it quickly caught on as a safety and comfort improvement in a wide range of industries. Sturtevant quickly became the father of the American fan industry, and he became rich. By 1867, his fan manufacturing plant in Jamaica Plain was the largest air-handling firm in the world. It was sold to Westinghouse in 1945.

were easy prey for Forster's salesmen, who followed the shills and took the orders.

Boston's Union Oyster House is said to be the first restaurant in America to offer free toothpicks, but soon they were dangling from lips everywhere, used not only for cleaning teeth but also as symbols of high fashion. A fad had begun, and now Forster could barely keep up with demand.

Soon after moving his enterprise to a larger mill near the Strong railroad depot, the narrow gauge Sandy River and Rangeley Lakes Railroad was groaning with carloads of toothpicks being carried to Farmington to catch the Maine Central Railroad for distribution across the country. Trucks hauling heavy wooden shipping boxes of toothpicks once collapsed the wire suspension bridge crossing the Sandy River. At the height of production, the company rolled out twenty million toothpicks a day, and the tiny town of Strong became the "toothpick capital of the world."

Forster's mills operated for more than one hundred years, making not only toothpicks but also a wide variety of wooden products including applicators, clothespins, ice cream spoons, candy sticks, cocktail forks, mustard paddles, tongue depressors, and more. It was only a matter of time before a surge in the use of dental floss made cavities in the toothpick market, and the production of most small wooden tools moved overseas. The last Forster mill closed in 2002.

"Hang your chemistry and electricity! If you want to make a pile of money, invent something that will enable these Europeans to cut each other's throats with greater facility."

10

The Devil's Paintbrush

On a June morning in 1890, a large crowd gathered on a hillside overlooking Lake Wassookeag in Dexter and gazed at a man seated on the rear leg of a tripod that held what appeared to be a small cannon. A pounding chatter erupted, and the lake was suddenly pocked with the splashes of countless bullets. The townspeople knew what they came to see, of course, but they left aghast at having witnessed the first United States demonstration of the most lethal weapon yet known to man.

The triggerman was fifty-year-old Hiram Maxim (1840-1916), returning home after earning worldwide fame as a master of the marvels of electricity, and, most especially, as the inventor of the first portable machine gun. His homecoming surely brought him great satisfaction, for when he left town many years before there was nothing to suggest he would ever amount to much. Born in the nearby town of Sangerville in 1840, he was the first of eight children in a poor family that had little to show of any promise—except for a remarkable strain of genius.

Whether from fatigue or philosophy, his mother, Harriet, was inclined to let her children run wild, and Isaac, his father, variously occupied himself as an occasional farmer, a clever mechanic, and an incurable tinkerer. Although the family never traveled far, Hiram later claimed they moved eleven times before he was ten, all within the Piscataquis River valley.

The first family move, from Sangerville, came when Hiram was only four, and it followed an uproar centered on the Millerites, a hybrid sect of Protestants of which Isaac was an avid member. The Millerites not only believed Jesus Christ was certain to return, but they also insisted they knew exactly when he was coming—October 23, 1844. When Jesus failed to show up, there was all hell to pay. The perplexed Millerite preacher was promptly run out of town, and Isaac and his growing brood of Maxims were not far behind.

After only five years of formal schooling, fourteen-year-old Hiram was apprenticed to a carriage-maker, first in Corinth, and then to another in Starks, learning to paint decorations on wooden dashboards while earning his keep by tending sheep. When the Civil War began in 1861, two of his brothers joined the Union Army, but Hiram declared himself a pacifist and moved to join his uncle's engineering firm in Fitchburg, Massachusetts.

The irony of his moral stance for peace was not lost on those who followed his later creation of a rapid-fire gun. It is sometimes said he made the gun as a deterrent, in the belief that no one would dare go to war with a country that possessed such a weapon. That notion was quickly proven wrong, of course, but in any case, it is more likely that his motive was money. That evidence is in his story of being in Vienna in 1882 and meeting an American who told him, "Hang your

Postcard image of Hiram Maxim with the Maxim gun.

chemistry and electricity! If you want to make a pile of money, invent something that will enable these Europeans to cut each other's throats with greater facility."

If Maxim wasn't a true pacifist, he was most certainly an atheist. He constantly railed against the "propaganda" of American Christian missionaries in China and once prepared a 400-page scrapbook for the prominent Chinese diplomat, Li Hung Chang, making the case that not every American was so silly as to be taken in by impossible miracles or to believe in ghosts, holy or not.

The move to Massachusetts was the first time he strayed outside of Maine, and it later prompted his wry observation that "Maine is the best state in the Union to emigrate from, and I had long wished to get out of it and go to some place where I could get more from my work."

Hiram Maxim, Louis Cassier and J. Bucknall Smith with a machine gun for Germany.

His every wish came true. By war's end, he had patented his first invention, an electric hair-curling iron (1866), and over the next several years created a veritable flood of original devices ranging from magno-electric machines to various oil, steam, and gas engines. When his experimentation in magnetic fields caused his pocket watch to slow, he promptly invented and patented a demagnetizing apparatus, and after a nearby furniture warehouse burned, he designed the nation's first fire sprinkler system capable of dousing a blaze and alerting the fire department.

It was an era of leaping progress in electrical science, and Maxim's knowledge and growing experience were precious. The transatlantic cable had been laid for ten years; Thomas Edison was designing power plants; and just two years before—in 1876—Alexander Graham Bell had spoken on the telephone with Thomas Watson. When the Electro-Dynamic Light Company of New York formed in 1878, Maxim was promptly recruited as chief engineer. He proceeded to design and install the first electric light system in any New York City building (Equitable Life, Broadway), and at the same time engaged Edison in a patent dispute over the incandescent light bulb. Maxim claimed the invention was his, and that when an employee made a false application for the patent, Edison uncovered the fraud and thus opened the way to claiming it for himself. Whatever the truth, Edison won the patent race by a matter of only days.

In 1881, Maxim moved to England to organize the company's European division, and promptly set about to build an efficient machine gun. It had been a dream of his since he was a young man when he and his father designed a prototype, but lacked the money to build it. Years had passed, and now he would have to follow other footsteps.

Maine Weaponry

Hiram Maxim was not the first Maine inventor to dabble in weaponry. In 1811, while working in his Portland tannery, John Hancock Hall (1781-1841) fashioned a breech-loading rifle.

Attracted by the efficiency of its uniform parts that enabled mass production and quick repairs, the U.S. Army adopted the Hall rifle as a replacement for the Harpers Ferry rifle-musket.

Just before the Civil War, Sumner-born attorney Lorenzo Dow (later mayor of Topeka, Kansas) invented the waterproof cartridge and went on to supervise its production for the Remington Arms Company. In 1871, Edward Very of Belfast, a Naval Academy graduate, invented the eponymously named Very Pistol for firing flares. Hiram Maxim's brother, Hudson, born in Orneville Township, was an atomic engineer and the inventor of the first smokeless gunpowder. In counterpoint to Hiram's noisy machine gun, his son, Hiram Percy Maxim of Wayne, later patented a noise-suppressing gun silencer. It came too late for his father, who suffered severe deafness from his work with explosives.

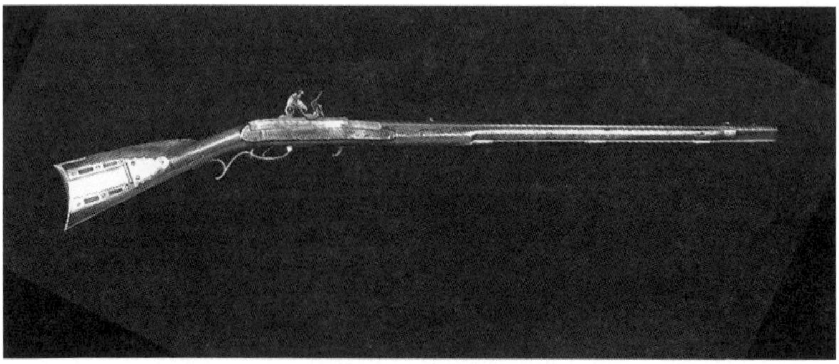

Extremely rare Model 1817 U.S. contract breechloading flintlock rifle from John H. Hall's first government contract for 100 firearms, which were sent to Fort Bellefontaine. Fewer than five examples are known to exist. Title: Model 1817 Hall U.S. Contract Breechloading Flintlock Rifle. Date 1817.

Richard Gatling had patented a rapid-fire gun in 1861, in time for use by the Union Army. The problem with Gatling's gun was that it took two men to make it work, was spring-loaded and hand-cranked, and awkward to move. Maxim wanted a one-man, automatic, and fully portable piece.

Hiram began his work by experimenting with several methods of machine gun operation, and finally settled on an action that used recoil energy to close the breech and compress a spring to prepare the gun for successive firing. He obtained a patent for the automatic self-loading and self-firing gun in the summer of 1883, and the first demonstration was given in England in October 1884.

The early models often jammed, and United States military leaders were at first wary of adopting it. They would soon change their minds, but in the meantime Maxim found a ready market in Europe, where there were wars aplenty. One of the first buyers was the peace-loving country of Switzerland, whose officials made the same false argument that such a destructive firearm would surely be a deterrent to aggression. (The same case would be used again to defend the coming generation of atomic weapons). Although the psychological impact of the Maxim gun may have saved lives in the late nineteenth century European occupation of Africa, its eventual use in World War I was devastating, and it became known worldwide as the devil's paintbrush.

The British Army adopted the gun in 1888, and it was first introduced into the United States arsenal ten years later, in time for the Spanish-American War. By World War I, every combatant had some version of Maxim's gun. The British Vickers machine gun was merely a revision of the Maxim, redesigned by Hiram himself, and the Russian Pulemyot was a Maxim copy. The gun that rained bullets accounted for much slaughter in the close-combat trenches of World War I, and in the world war that followed, Allied and Axis powers together had more than fifty versions of Maxim's gun along every front of the war.

Comfortably ensconced in a palatial country house in West Norwood, and close by his London factory and his European customers, Maxim became a naturalized British subject in 1899. A year later, Queen Victoria announced her intention to grant him knighthood for his many special achievements. Her Majesty died before she could carry it out, but her son, King Edward VII, invested Sir Hiram with his title in 1901.

Sir Hiram died in London in 1916. He was 77. At the time of his death, he held 221 U.S. patents. Despite his many inventive contributions, the mists of time might have assigned his name to obscurity were it not for his creation of a weapon that could spew a dozen deadly rounds in a single second.

Harry Oakes

Sir Harry Oakes

In 1939, another son of Sangerville became a British hereditary dignitary when King George VI made millionaire Harry Oakes (1874-1943) a baronet. Oakes joined the Klondike Gold Rush after leaving Bowdoin College, and in 1911 registered a claim to what became Lake Shore Mine in Northern Ontario, Canada, one of the most productive gold mines in the Americas. Eager to escape United States taxes, he became a British citizen in 1935 and made his permanent residence in The Bahamas while keeping his summer home in Bar Harbor. He was made a baronet (a step above a knight) in recognition of his philanthropy in The Bahamas, Canada, and Britain. He was mysteriously murdered at his Nassau estate in 1943. With claim as the birthplace of both Sir Hiram and Sir Harry, it can be said that no small town in America is better represented in the line of those titled by the Crown than Sangerville.

Wellington Kidder's genius continued, and he went on to invent a consecutive numbering machine that was adopted by the government for numbering bank notes and by ticket sellers everywhere.

11

The Three R's

The years following the Civil War brought the reunited states a period of astonishing change and growth, unmatched before or since. Over rivers and rails, the migration west accelerated on the power of steam. Whole cities everywhere grew up around new factories. Taken together with the new marvel of electricity, it seemed the Industrial Revolution might never end.

With the burgeoning growth of manufacturing and sales, it was nearly impossible to keep track of things by hand, and the nation's bookkeepers were desperate to find a better way. Credit for the design of a mechanical calculating machine rightfully belongs to British mathematician Charles Babbage (1791-1871), who began his work on the design for a "difference" machine in 1819. Despite the support of the British government, metalworking techniques of the time could not produce the precise parts, including gears, needed for such a machine, and the project was abandoned in 1833. Babbage never built the machine of his own design.

Over the next many years, Babbage and others produced limited-capacity examples of the calculator, but it wasn't until 1872 that the son of a Maine shipbuilder built the first full-scale model and began leading the nation's number crunchers out of the swamp.

George Grant

George Barnard Grant (1849-1917) of the Farmingdale section of Gardiner attended Bridgton Academy and went on to Dartmouth where he became interested in calculators while working on mathematical tables for a construction excavation. It wasn't until he transferred to Harvard's Lawrence Scientific School in 1869 that he became aware of Babbage's work and set out to build the first full-scale machine. Following his graduation in 1873, the Boston Thursday Club helped raised the funds for the project, completed in 1876.

Grant's Calculating Engine at the Centinial Exhibition in 1876.

Consisting of some fifteen thousand parts, the Grant Difference Engine was the size of a small stable and weighed nearly a ton. It had the remarkable capacity to calculate twenty-digit numbers and sixth-order differences, and when it was demonstrated alongside one of Grant's smaller general purpose calculators at the Philadelphia Centennial Exposition in 1876, it was hailed as a "thinking machine." A century later, models displayed at the Smithsonian Institution and the University of Pennsylvania, are cited as examples of the first early steps into the computer age.

Until the end of the century, Grant continued to patent and produce several models of calculators at his own company in Lexington, Massachusetts, but his greatest success and fame came from his innovative work in manufacturing gears, a puzzle that had eluded his fellow pioneers in the realm of calculators.

When he began work on his full-fledged difference machine, he opened a gear-cutting machine shop in Charlestown, Massachusetts, where he cut precision gears that could not be found anywhere else. Using a "hob," a tiny device of his own invention, he was able to make gears of unmatched accuracy, and was soon receiving orders from around the world. To be nearer the centers of shipping and sales, Grant Gear Works soon moved to Boston and later opened branches in Philadelphia and Cleveland.

Wellington Kidder

While the bookkeepers were well on the road to the world of high technology, the tools of writers and printers hadn't changed much for four hundred years until Norridgewock-born Wellington Kidder (1853-1924) began making a number of improvements for both printing presses and typewriters.

Wellington Parker Kidder

In 1874, Kidder built the first web form of an offset press that enabled the use of a continuous roll of paper. Within ten years he patented additional press improvements, and, taken together, his inventions increased the speed of high-volume printing fourfold. His genius continued, and he went on to invent a consecutive numbering machine that was adopted by the government for numbering bank notes and by ticket sellers everywhere.

Wellington Kidder was given a leg up in his press inventions by the work of North Berwick native Thomas Jefferson Mayall (1828-1888), who had earlier invented a cylinder printing machine, first used to mass produce wallpaper. In his lifetime, Mayall would claim some 200 U.S. patents for endeavors ranging from a steam-driven cannon to a railroad drawbridge.

Although Kidder's enhancement of presses brought him notoriety, his popular fame came from his improvements of the writing machine, a constantly cloned device first commercially available in 1874. His Franklin model, invented in 1887, was the first "visible" typewriter, which allowed the operator to watch the key strike the paper. He later produced the Canadian Wellington typewriter, and in 1895, the first noiseless typewriter, later purchased and adopted by Remington.

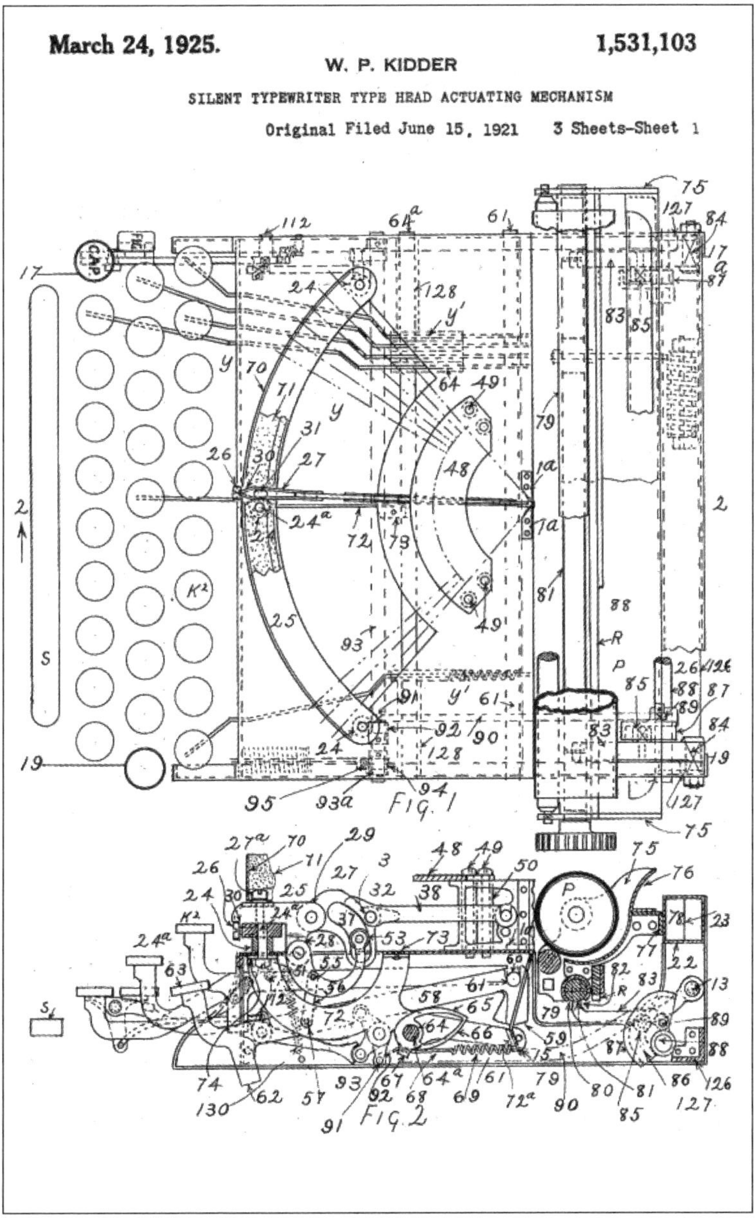

Illustration for silent typewriter patent application, 1925.

Despite Leonard Atwood's success with railroads, his broad-ranged tinkering never stopped, and in 1907, he patented the first fishing reel with adjustable tension (drag).

12

The Gilded Age

From the end of Reconstruction to the beginning of the new century, the Gilded Age brought the most rapid economic and social change in the nation's history. American entrepreneurs scrambled to seize new technologies and ideas, and their work helped to carry the nation to the very pinnacle of the world's industrial powers. Prominent in the queue of eager patent seekers were a number of Maine inventors who brought forth an eclectic array of creations ranging from sardine cans to carpet seaming, and from washing machines to elevators.

Among the many Maine inventors of the age, none had a more direct positive impact upon the unique affairs of his home state than Charles Henry Emery (1826-1925), a Portland native who worked in the wholesale beef business while earning a theological degree from Westbrook Seminary before changing his life's path. After training at Shaw's Business College in Portland, he moved to Chicago in 1881 to work for the Libby McNeil & Libby food canning company, and while serving as superintendent of the firm's can-making division, he developed and patented a machine that soldered the tops of tins of meat.

Previously, the tops had been soldered by hand, a painstakingly slow and costly process. Although meat tins were of a rectangular shape in order to extract the product for slicing, Emery's machine worked on round tins as well, and it revolutionized the canning industry. (Among the Libby products was corned beef hash, an old New England recipe for leftovers, provided by Emery's wife.) Nowhere was the soldering invention more welcomed than in Downeast Maine, where sardines were packed in rectangular tins, albeit not for easy extraction but for easy insertion.

In 1890, Emery returned to his native state to install soldering machines in some forty Eastport canning houses of the Seacoast Canning Company. While the job was being done, he opened his own Independent Canning Company along that same stretch of coast.

The demand for Emery and his invention never stopped. He later returned to the Libby firm and was sent to Alaska to pack salmon, then to New Jersey to seal cans of tomatoes. Before he retired in 1924, he purchased citrus groves in California and Florida, providing juices to pack in his own soldered cans.

Starrett and Poland

Like Emery, many inventors sought ways to improve the efficiency in their own trades. In 1879, Leroy Starrett (1836-1922), a carpenter in China, patented a combination square and ruler with a sliding head. The tool, essentially unchanged, is still routinely found in toolboxes of woodworkers everywhere. And, as an employee of Bates Mill in Lewiston, John Poland of Lovell led the way in the creation of the modern washing machine by inventing the centrifugal wringer (1884),

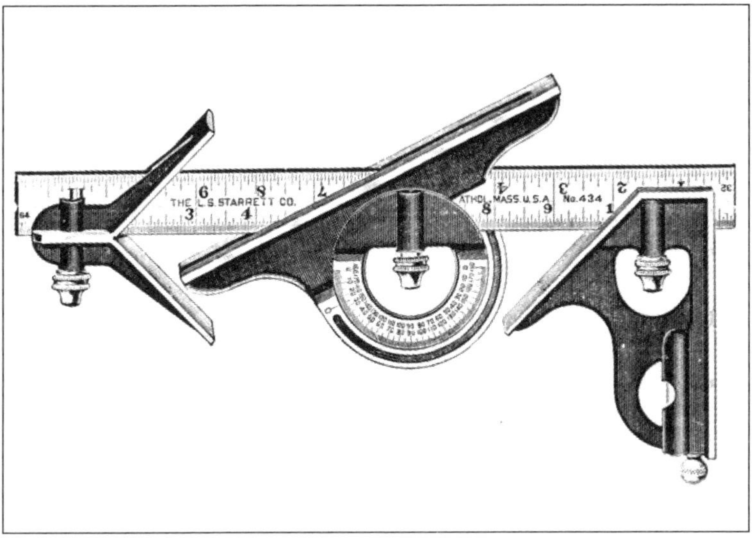

Illustration of a combination square, including a center finder head (left), protractor head (center), and a standard head (right).

an automatic reverse-spin mechanism (1885), and the mangle (two-roll press) wringer (1886).

Edward Allen

Others were inspired to invent gadgets long before they had established careers. Edward Beecher Allen of Naples was only thirteen when he patented a can opener, 14 when he devised an improved loom shuttle. Doubtless confident of his future as an inventor, he dropped out of school the following year and became an apprentice at the Portland Company, a maker of locomotives, before joining his father's shoe company in the same city. Here, he invented a machine for putting heels on shoes, which he sold to the McKay Heel Company in Lawrence, Massachusetts, where he became a partner.

Leonard Atwood circa 1930
Courtesy of the Farmington Historical Society and Jeffrey Knapp.

His meandering was not over. In 1888, after opening his own repair shop in Portland, he invented a buttonhole clamp that flagged his future career. He took the clamp with him to the Singer Sewing Machine Company in Elizabethport, New Jersey, before moving to the firm's headquarters in Bridgeport, Connecticut, and was given his own laboratory where he worked out many industrial sewing improvements.

Over the next several years, Allen collected more than two hundred sewing related patents, including a machine that seamed carpets, a job that previously could only be done by hand. In something of an advertising stunt, he replaced all of the carpets at New York's Metropolitan Opera House in a single day.

Leonard Atwood

Another vagabond Maine inventor was Leonard Atwood (1845-1930) of Farmington Falls. He mustered out of the Union Navy when he was barely twenty and settled in Pennsylvania where he promptly patented an improvement for a drilling rig that he first used near Titusville, tapping into oil at five hundred feet. The man with the Midas touch then proceeded to build the world's first oil pipeline (unpatented) and take a job with John D. Rockefeller Sr., who was then on his way to establishing Standard Oil.

Within two years, Atwood patented the vertical railway, a pulley system that allowed loading platforms to be moved up and down, and began installing these "elevators" not only in the freight warehouses of the New York Central Railroad, but soon also in major hotels in New York City and in Washington, D.C., where he installed the first elevator in the nation's Capitol. Already credited as the "Father of the

Skyscraper," the venturesome Atwood sold his elevator patent rights to Elisha Otis (1811-1861), who was already busily building an empire.

Otis was a Vermont native and another restless man. He began his career as a doll maker and set to work inventing a train safety brake on the side. He later moved to New York to establish his own bedstead company where in 1851 he constructed a hoisting platform of his own design to which he added a version of his safety brake. Two years later, he stood before a large crowd on one of his platforms in the Crystal Palace at the New York World's Fair and invited an associate to cut the hoisting rope with an axe. The platform fell only an inch or two before the brakes grabbed. After buying Leonard Atwood's plan for a vertical railway, Otis had all he needed to build an elevator people would dare to ride.

Meanwhile, well-versed in the mechanics of the rails, Atwood turned to making electric rail cars and in the early 1880s, moved back to Farmington Falls and built the Franklin Pulp Mill, later part of International Paper Company. He didn't stay home long. His wealth and reputation as a master mechanic led him to become a partner of the Coast Railway Company, chartered to build a rail line to span the Nova Scotia peninsula from Yarmouth to Lockeport. By the time that work was done, railways were beginning to convert to the wider and stronger standard gauge, and Atwood returned again to Maine to take over the Wiscasset & Quebec Railroad, which in 1898 he reorganized as the standard gauge Wiscasset, Waterville & Farmington Railroad Company.

Despite his success with railroads, his broad-ranged tinkering never stopped, and in 1907, he patented the first fishing reel with adjustable tension (drag). The smooth spindle bearings and transparent sides made it the finest of its kind, but it was so exquisite and complicated

Elisha Otis demonstrates his free-fall prevention mechanism, Crystal Palace, 1854. Otis purchased Leonard Atwood's plan for a vertical railway to create an elevator people would dare to ride.

that it had to be priced above the market and it was impossible to sell. Even so, it became a prototype of the improved and less expensive reels that followed.

Frank Bunker Gilbreth

With all the hustle and bustle of fast-growing industries, there was a growing need for organization and efficiency. Frank Bunker Gilbreth (1868-1924) of Fairfield was an international expert. A graduate of Boston English High School, he declined an offer of admission to Massachusetts Institute of Technology and began his career as a bricklayer's apprentice. In 1895, he became a self-employed general contractor, building mills, dams, and power plants around the United States and in Europe. Along the way, he invented a number of

Frank Bunker Gilbreth Sr.

Cheaper By the Dozen

The Gilbreths and their family of twelve children were immortalized in the 1950s autobiographical book, *Cheaper by the Dozen,* written by his son, Frank, and daughter, Ernestine. The book, set in the 1920s, is based upon the family's life in Montclair, New Jersey. It was later made into a Hollywood film starring Clifton Webb, Myrna Loy, and Jeanne Crain.

building tools and machines including a safety scaffold for bricklayers, conveyors, and an improved concrete mixer.

The complexity of his many operations led him to formulate the first cost-plus-fixed sum contract and to develop a number of systems to reduce waste, monitor work progress, and improve the productivity of his workers. He and his wife, Lillian Moller Gilbreth, a psychologist, conducted time and motion studies and built a reputation as efficiency experts, and were often called upon to present their findings at colleges and universities and at meetings of leading scientific and industrial societies.

Dubbed "Flying Teapots," the Stanley's first models had wooden bodies mounted on tube steel frames and were powered by a boiler set over a gasoline burner beneath the driver's seat.

13

Stanley Steamer

By the end of the century, the face of America had changed in countless ways, and although iron rails now connected towns across the nation, much of the country still relied on the trusty horse for toil at home and local transportation. As the clock ticked toward the twentieth century, most every country barn still held one or more of the obliging beasts, stables and carriage houses continued to be built alongside rich men's new mansions, and hitching posts and watering troughs were still tended in the centers of every town.

All of that was about to change in a historical second.

The first hint came in 1859 when Belgian engineer Étienne Lenoir developed the first internal combustion engine, and inventors around the world began to ponder the endless ways it could be used. A German, Nickolas Otto, improved the engine in 1876, and in 1885 his countryman Karl Benz built the world's first self-propelled motor wagon. Ten years later, the American genius Henry Ford assembled a "quadricycle" in Dearborn, Michigan, and the country quickly got

Frances and Freelan Stanley riding in one of their famous steam-powered cars.
Courtesy of the Stanley Museum.

down to the business of widening the roads along the routes of former cart paths to make room for new horseless carriages that would soon be sending real horses and carriages into the ditches.

At the outset, there were some who felt the gasoline engine was too complex and dangerous for an automobile, and that it was more practical and efficient to use the proven power of steam. Among them were identical twins Francis (1849-1918) and Freelan Stanley (1849-1940) of Kingfield.

Born in 1849, the Stanleys were already successful businessmen when they set about to build an automobile. Their younger sister,

Chansonetta Stanley Emmons (1858-1937), was an accomplished photographer, acclaimed as one of only a few who recorded rural domestic scenes, often on brilliantly hand-colored lantern slides. Freelan had made a small fortune helping his sister with his 1886 patented refinement of the dry plate photography process (allowing for glass negative plates to be stored both before and after exposure), and in

---

Fred Marriott in his "Rocket" Stanley Steamer. Circa 1906 and 1907.

The Need for Speed

Anxious for public notice, in 1906 the Stanley brothers entered a "Rocket" model of their steamer in the 1906 Daytona Beach Road Course. Driven by Fred Marriott, the car lived up to its name, traveling the astonishing speed of 127 mph. The car covered a mile in 28.2 seconds, a world record that held for the next five years.

1888, he and his brother partnered in the Stanley Dry Plate Company, first in Kingfield and then in an old bicycle factory in Watertown, Massachusetts, near their growing sphere of customers.

The brothers began to build their automobile in Watertown, but soon moved to a larger space in Newton, where the first Stanley Steamer was completed in 1897. Over the next two years the factory produced two hundred cars, becoming the first company in the country to engage in the commercial sale of automobiles. Encouraged by their automotive success, the Stanleys sold the dry-plate patent to George Eastman (who used it to help build the Eastman Kodak empire) and turned full time to building cars.

Dubbed "Flying Teapots," their first automobile models had wooden bodies mounted on tube steel frames and were powered by a boiler set over a gasoline burner beneath the driver's seat. Without transmission or clutch, the engine had side-by-side cylinders that powered the rear axle through a differential gear at the end of a chain-driven crankshaft. The auto was capable of short sprints of one hundred horsepower or more, and as a publicity stunt, in 1899 Freelan and his wife, Flora, drove a Stanley car to the top of Mount Washington. The 7.6-mile trip up took more than two hours. With squealing brakes, it came down the mountain a great deal faster.

In 1899, not satisfied with their early design, the Stanleys sold the rights to the publisher of *Cosmopolitan*, John Walker, who became the maker of the hand-built steam Locomobile. The twin brothers opened the Stanley Motor Carriage Company in Newton and began to develop a completely new design.

All the while, competition grew almost overnight. In 1901, Ransom Olds began producing his internal combustion engine cars on a crude assembly line in Lansing, Michigan, and in 1903 Henry Ford opened his first plant seventy-five miles away in Dearborn.

By 1908, Henry Ford was making an affordable car for the average man, and in 1913, using a full-chassis assembly line for the first time, the company completed a single Model-T every two hours and thirty minutes. Over the next ten years, fifteen million Ford cars were built and sold.

To combat the popularity of the Ford, the Stanley brothers developed a new model, adding condensers, placing the steam boiler in the front, and gearing the twin-cylinder engines directly to the rear axle. At the same time, they launched an advertising campaign that derided their competition for their internal "explosion" engine and bragged that steam engines never blew up.

While there was truth to their claims, no amount of advertising could overcome the fact that new car owners were, as a lot, in a hurry. They wanted to get in and drive away, and had no patience to wait for the Stanley to warm up and make a head of steam. Electric starters of internal combustion engines had first come on the market in 1912, and while later models included a crank "just in case," drivers loved the floor-mounted starter button.

The steamers were plainly doomed by 1917, when the Stanleys managed to hand-build five hundred cars—the highest production ever—while Ford churned out 735,000 Tin Lizzies. To further widen the edge, the cost of a Stanley sedan was nearly $4,000. A Model-T was less than $500.

With some irony, Francis Stanley died in an automobile accident in 1918, and his bereaved brother sold his interests in the company, which sputtered for a few years before closing its doors for good in 1924.

For harvesting wood, the Lombard Hauler was as significant as Eli Whitney's cotton gin had been for reaping cotton a century before, and its impact reached far beyond Maine.

14

The Log Hauler

On a summer afternoon in 1899, Alvin Lombard (1856-1937) boarded a streetcar in Fairfield, bound for his majestic new home on Elm Street in Waterville. By chance, his seatmate was E.J. Lawrence. The two were not strangers, and both men were well known around the state. Only forty-three, Lombard had already earned fame and fortune from his various inventions, and Lawrence, age sixty-six, was principal owner of Lawrence, Newhall & Page, a prosperous Kennebec River lumber debarking firm soon to give a name to both Shawmut Village and its manufacturing company backed by the Shawmut Bank of Boston

At the time, Maine's lumbering business was at a crossroads. In the early years, oxen were used to move the giant pines to the rivers for the spring drives, but by the turn of the century, the virgin pines had all but disappeared, and draft horses were used to harvest the thousands of acres of valuable maple, birch, beech, and ash that had been left behind. Here and there, narrow gauge railroads poked into the

Lombard Log Hauler, early 1900s

woods to help retrieve the trees, but the trains were of little use when the harvesting sites were far from the rigid tracks, and the hardwoods could not easily be floated down the rivers. A better way had to be found to carry the timber overland in the deep snows of winter. As they rode along on the streetcar, Lawrence bemoaned the cost and cruelty of using horses, and asked if Lombard thought it might be possible to build a machine that could replace them.

Two days after their chance meeting, Lombard appeared in Lawrence's Fairfield office with a carved wooden scale model of a revolutionary new logging machine he said could haul itself along on continuous treads. The two quickly agreed to build a prototype at the nearby Waterville Iron Works, and on May 4, 1901, U.S. Patent

#674,737 was issued for the Lombard Log Hauler, arguably the most significant invention ever to come from the State of Maine.

Each man then went his own way, but each made enormous contributions to their time and place. Lawrence found new riches with Lombard's machine, and quickly began to give them back. Within a year of employing the Lombard invention, he built his hometown its first free public library, and in 1907 gave a new high school, dedicated in his name.

Like many geniuses of his era, Lombard had taken an unlikely road to success. Born in the Penobscot County town of Springfield in 1856, he was poor and unschooled, and at the age of eight began working at the family shingle mill in nearby Lincoln. Stoutly built, inquisitive, and energetic, he quickly mastered every woodland task from lumberjack to river driver and from stacker to mill sawyer. Along the way, he astonished others with his creations. As a young boy, he once built a miniature sawmill powered by a stream near the family farm, and demonstrated its worth by cutting cucumbers into boards. Later, he found commercial success when he invented a device for rossing (de-barking) pulpwood, and an apparatus that separated knots and sawdust from ground pulp.

Over his long career, the invention of which he was most proud was not the ubiquitous log hauler, but an automatic mechanical device he designed that maintained the speed and power of water turbines. The 1893 innovation found immediate and widespread uses, and he manufactured and sold his "governor" for six years before deciding to get out of commercial business altogether. He sold his patent and the business and was biding his time between his Waterville home (with

---

O.C. Johnson and the First Snow Machine

Steered by skis and driven by continuous belts, today's ubiquitous snowmobiles are but miniature versions of Lombard's creation. Doubtlessly taking his cue from Lombard, another Waterville man, O. C. Johnson, is said to have built one of the first snow machines in 1909. It was ten feet long and powered by a "one lung" engine. The commercial snowmobile did not come along until 1954 when a Minnesota farm equipment company named Polaris sold its first machine after extensive testing in the hills of Northern Maine.

---

a workshop in the basement) and his country home in Vassalboro when he built the log hauler.

By 1903, demand for the new hauler far exceeded the ability to turn them out, and the new Lombard Traction Company moved out of the Iron Works and up College Avenue to the Fairfield line.

The time for the giant machines was ripe. The development of internal combustion engines was moving apace, and other gasoline and steam log hauler designs were in the works. James and Ira Peavey of Bangor tried using a sled that advanced on a giant screw, but it didn't work well enough to replicate. For his rig, Lombard chose the well-established steam engine with an upright boiler that drove engines, front and back. An outsized bobsled in the front was geared to steer.

All of that was almost ordinary, but what made his machine unique was its method of propulsion. A continuous belt of hinged steel lags (treads) was fitted over two pairs of geared wheels, allowing the heavy machine to pull itself along on a rolling carpet of steel, over stumpage and snowdrifts, like a caterpillar.

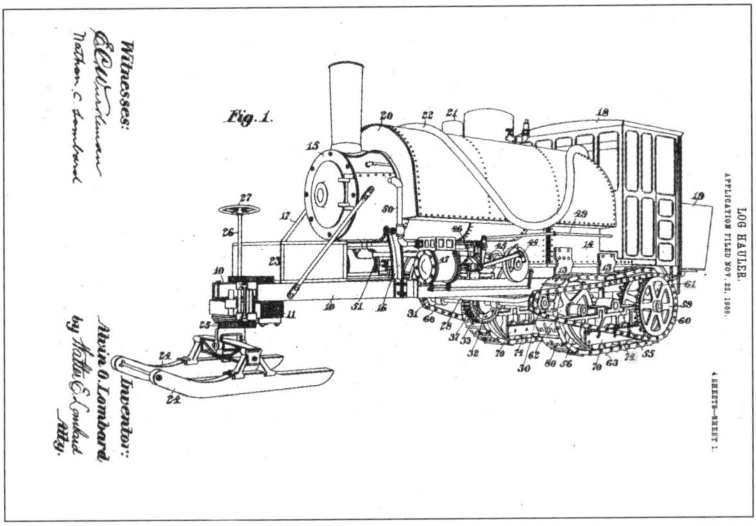

Illustration to accompany patent application for Lombard Steam Log Hauler, 1907.

For harvesting wood, the Lombard Log Hauler was as significant as Whitney's cotton gin had been for reaping cotton a century before, and its impact reached far beyond Maine. European and United States military leaders quickly adopted the tread for tanks, and manufacturers, including the Illinois firm that took the Caterpillar name, began using them for farm tractors and heavy equipment.

Having made eighty-three steam-powered haulers by 1917, Lombard began experimenting with gasoline engines, but they lacked the sheer power of the steam versions, which could each carry fifty tons of logs or more. His first experiment with a diesel engine in 1934 was the last hauler he ever built. By then, lumbermen were turning to the use of diesel-powered trucks, and while the thirty-year reign of the Lombard Tractor was over, the caterpillar tread was, true to its name, never-ending.

Martin Keyes was set to go, but when he applied for a patent he discovered a co-worker had stolen his idea and already held the license.

15

Paper Plates

Like Alvin Lombard, Martin Keyes (1850-1914) was blessed with an inquisitive and clever mind, but unlike his burly tractor-making neighbor, Keyes was a diminutive and fastidious man. He kept a diary every day, and it was his thorough way that led him to claim an invention that established one of Maine's most successful international industries.

As a young boy, Keyes worked in his father's sawmill before venturing on his own as a maker of sleighs and carriages in his hometown of Lempster, New Hampshire. He later took a job at a veneer mill in upstate New York, where he is said to have been struck by the idea of making disposable plates while watching workers eat their lunches on thin scrap slices of maple veneer.

Working on his own, he first tried steaming and shaping thin veneer into plates, but they were too brittle, and he gave up the quest until he moved to Maine, where he became superintendent of the

Women packing plates circa 1909 at Keyes Fiber Company mill (now Huhtamaki), in Waterville.
Courtesy of City of Waterville.

Indurated Fibre Company in North Gorham. Here, with the support of his employer, he turned to the idea of making his plates from molded wood pulp. The process required many steps, and it took Keyes several years of experimenting before he had it right.

Finally, in 1902 he took his drawings to Edmund Sprague at Portland Iron Works, who agreed to build a prototype. With flailing arms that hissed and groaned as they rotated through a process of dipping, drying, ejecting, and packaging, the finished apparatus resembled the complicated contrivances of the cartoonist Rube Goldberg.

Above: Keyes Fiber Company mill, 1910 (now Huhtamaki) in Waterville. Left: Martin Keyes, founder of Keyes Fiber Company and inventor of molded fiber technology in 1903.
Courtesy of City of Waterville.

Keyes was set to go, but when he applied for a patent he discovered a co-worker had stolen his idea and already held the license. His only recourse was through the courts, and when he was able to get his meticulous diary admitted into evidence, the fraud was exposed.

With patent in hand, Keyes rented a section of a new pulp mill in Shawmut, took on a local man named Bert Williamson as his assistant, and on November 2, 1903, the Keyes Fibre Company began churning out fifty thousand plates a day. By the time of the first shipment (June 24, 1904) the mill already had its more common and enduring name: The Pie Plate.

The inventor's troubles weren't over. Industry challengers were undercutting his prices and finding a steady supply of pulp was near impossible. He closed for several months in 1905, then reopened and was ironically saved by the great San Francisco earthquake and fire of 1906, which brought a factory-saving order of an entire carload of paper plates.

Just as business was beginning to improve, the owners of the Shawmut pulp mill decided to close. Keyes considered moving his plant to Portland, but on a visit to that city he was dismayed to find many drunken workers. A firm prohibitionist, he decided to move to Waterville instead. In 1907, he chose a site immediately north of Lombard's tractor factory, and the following year began to build a single-story brick building that would eventually stretch for more than a quarter mile along the town lines. The plant was soon making more than a million plates a year with sales bolstered by its first consumer product, a "picnic package" of varied dishes.

Keyes died in 1917, but his son-in-law, George Averill, kept the plant in family hands until 1927 when it was subsumed by majority shareholders in a corporate takeover. Despite the change, the Keyes name continued and the company grew with the issuance of each new product line, first KYS-ITE, a fibrous plastic; and then Chinet, the first leak-resistant line of disposable paper dinnerware.

The Finnish company Huhtamaki, worldwide maker of more than four hundred wood pulp products, now owns Martin Keyes' creation, and the Waterville plant continues as one of Maine's oldest manufacturers.

Like the Wright brothers, Charles Lamson also ran a bicycle shop, which he too eventually ignored as he chased his dream of flying.

16

Lamson's Kite

By 1899, Orville and Wilbur Wright had abandoned their bicycle shop in Dayton, Ohio, and were working full time in full pursuit of their passion to build an aeroplane. There was little time to waste. Increasingly, newspapers across the country were ablaze with headlines touting the achievements of other pioneering inventors who had taken on the battle with gravity and were striving to become the first to fly a powered aircraft. A Maine man came very close to winning the race.

Born in Augusta, Charles H. Lamson (1847-1930) was a successful watchmaker and jeweler in Portland when he took up a passion for building kites. Like the Wright brothers, he also ran a bicycle shop, which he too eventually ignored as he chased his dream of flying.

Lamson's kites were not simple toys, but large flying machines capable of lifting a man. His 1896 version was based upon the box kite, first designed three years earlier by the Australian engineer Lawrence Hargrave. With two parallel rectangular boxes attached in the middle

AIR SHIP ON THE GROUND.

THE KITE-AIRSHIP.

Illustration for Charles H. Lamson's Kite–Airship, 1896.

with struts that held a seat, "The Lamson Airship" was first launched in Portland on August 20, 1896.

As a caution, on the first flight Lamson placed a 150-pound bag of sand in the seat instead of a man, and as some 1,500 spectators gaped in amazement, fifteen men hauled the kite onto the field and played out four hundred feet of rope behind it. On Lamson's signal, the small ship was released to the wind and slowly rose into the air. It had reached an altitude of about six hundred feet when the tether snapped. Onlookers gasped, expecting a crash, but the kite and its bag of sand settled slowly back to ground, unscathed.

An article in the next day's issue of the *New York Sun* hailed the achievement and noted that it had been "the largest kite or airship ever floated," and that it had carried "the heaviest weight to the greatest altitude" on record. The article went on to quote the aeronautical authority W. A. Eddy who called the feat "the greatest step toward solving the problem of aerial navigation of the age."

Impressed by Lamson's work, the United States Congress appropriated money to further experiment with his kites at the Blue Hills Meteorological Station in Milton, Massachusetts, and a year later Lamson took to the skies himself on two kite trips in Connecticut, each time ascending to fifty feet or so and staying up for about a half-hour.

His ventures prompted the *Waterbury Democrat* (August 11, 1897) to applaud the achievement even while lamenting, "the weak point in his invention is that while he can raise himself with his kite at will, he cannot lower himself. But this defect he hopes soon to remedy." Nonetheless, the paper called the achievement "a distinct step forward towards the accomplishment of practical navigation."

Lamson promptly solved the landing problem by hinging the rear wings and providing a lever for the operator to use in maneuvering. Other changes made it possible for the driver to take off while seated in the cockpit, whereas earlier versions had required the pilot to climb up rope and pulleys after the kite was aloft. He patented his design for the tilting wings and went on to build gliders for others, including the Wright brothers, who tested the Lamson Airship near Kitty Hawk, North Carolina, before building one of their own.

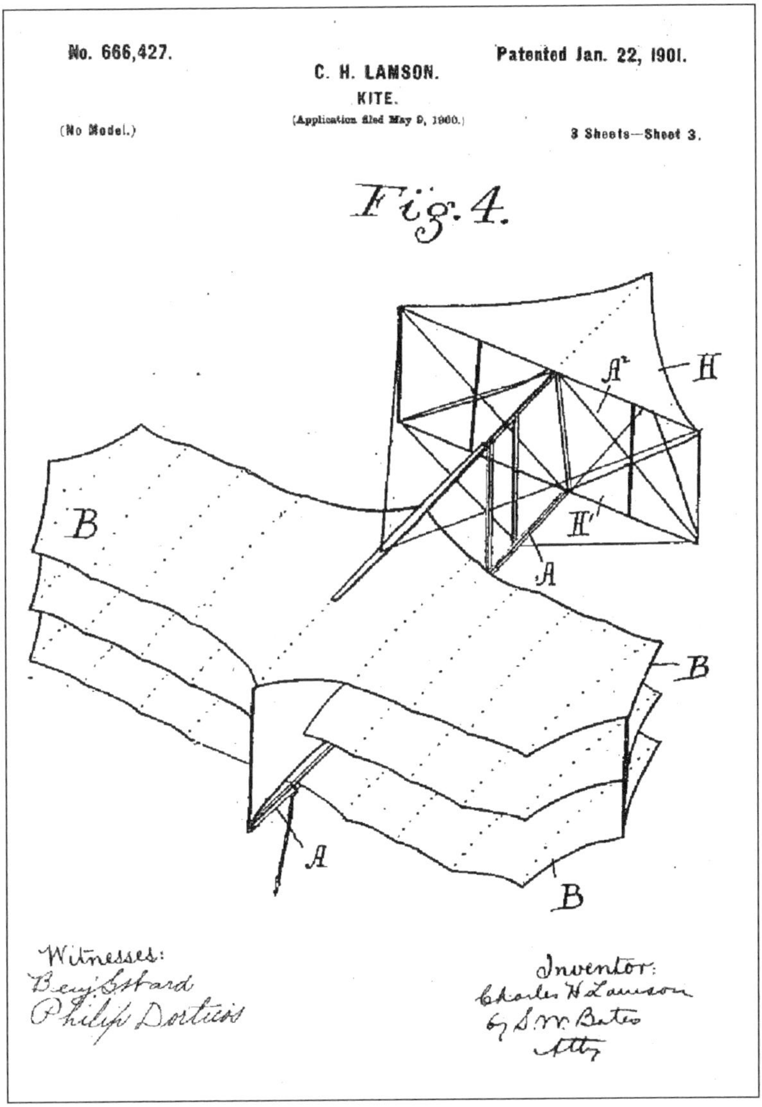

Illustration to accompany patent application for Charles Lamson's Kite, 1901.

After a single term at Hebron Academy, Leon Leonwood Bean turned to the Maine woods for his education and soon became an inveterate hunter, trapper and fisherman.

17

Bean's Boots

Leather had been used for making outdoor shoes since man first began to walk upright, and the world of footwear didn't change much until the Middle Ages when the Gauls attached wooden soles and called them galoshes. Generations of feet suffered, and another four centuries passed before cobblers thought to take Charles Goodyear's discovery and mold boots made of rubber. They weren't much good, either. The rubber often punctured, letting the boots take in water that promptly froze around the toes, and worse than that, the new galoshes were fastened with metal clips that bent or broke in the wrestling required to pull on the boots. Weary of wet feet, frostbite, and bloody knuckles, by the twentieth century, outdoor people were overripe for something new.

As with many other successful entrepreneurs, the early life of Leon Leonwood Bean (1872-1967) was pocked by tragedy and hardship. Born in the tiny town of Greenwood in 1872, he was orphaned at the age of twelve when both of his parents died within days of each other. Leon and his five siblings were sent to live with relatives in South

Paris, where his love of the outdoors soon outdistanced any fondness for classrooms. After a single term at Hebron Academy, he turned to the Maine woods for his education and soon became an inveterate hunter, trapper, and fisherman.

Although his disdain for cold, wet feet has become part of the lore, it is likely he was looking out for more than his own comfort when in 1911, at age thirty-nine, he designed a new kind of waterproof boot with rubber bottoms and laced, soft-leather tops. The following year he hired a cobbler to make a few pairs and mailed a crude flyer to all licensed Maine hunters, touting his new hunting boot and guaranteeing satisfaction.

His warranty was promptly tested as most of the first one hundred pairs were defective. It was a crushing blow to a man who dreamed of making a fortune, but he promptly sent each customer a refund. What might have been a business disaster turned out to be a boon to L.L. Bean, who had firmly established his reputation as a merchant who truly would sell only quality goods and satisfy every customer. Thereafter, the company never sold an item that wasn't first tested for quality and endurance, and the refund policy was underscored with every sale.

Following his initial hiccup in 1911, Bean procured a four hundred dollar loan and headed to Boston where the United States Rubber Company helped him make improved boot bottoms with a chain pattern. In 1917, he hired a few skilled cutters and sewers and moved his tiny operation out of the basement of his brother's Freeport apparel shop into upstairs space over the post office on Main Street. It was a most convenient location. The U.S. Postal Service had recently instituted domestic parcel post service, and his brother was the town

Satisfaction Guaranteed

Unlike most other retailers, the Bean guarantee of satisfaction was exercised at the discretion of the customers, not the company. The bold policy continued for nearly a century, but it was increasingly abused by customers who did not share Leon Bean's standard of integrity. Too many customers were maintaining wardrobes by returning goods that were damaged by abuse or simply worn out. In 2018, the company said it would guarantee what it sold for a year, unless defective, and would require proof of purchase for returns. It seemed only fair.

Above: A young Leon Leonwood Bean. *Courtesy of Digicommons.*
A photo of Bean boots.

postmaster. With boot making and mailing all in one place, it wasn't long before the tiny post office was overwhelmed by a tide of boots coming down the stairs.

The following year, he patented his creation with both the United States and Canada. Although it was registered as the Maine Hunting Shoe, customers never stopped calling them Bean Boots. As his new business got underway, Bean was usually on hand to sell his boots in person, all the while sharing his enthusiasm and wisdom about the outdoors. Gradually, the Freeport store and its famous catalogue began to offer more and more items of sporting equipment and outdoor apparel.

By 1934, the Bean factory covered 13,000 square feet and the catalog had grown to fifty pages. Leon died in 1967 at the age of ninety-four, and his grandson, Leon Gorman became his successor. The fascinating story of the company's remarkable growth to become a worldwide supplier of outdoor gear is an American classic, and its refinement of the catalog business and sales by mail are the topics of textbooks.

Unlike the coming and going fads of other shoe and boot styles, the popularity of the Bean Boot has never stopped rising, and the company still struggles to keep up with the demand from customers who range from serious outdoor people to fashion-conscious college students who rarely venture into the woods, but are eager to display their rugged Maine independence.

Today, just outside the never-locked doors of the flagship store in Freeport, a giant model of the Bean Boot (size 410) stands in testimony to the man whose inventive ingenuity made him famous and whose business integrity made him a legend.

Dozens of Maine tinkerers rode on that great wave, beginning with Joseph Stearns of Weld, who was working as a telegraph operator for the Boston Fire Alarm Company in 1868 when he developed and patented the duplex system of telegraphy that allowed simultaneous, two-way communication.

18

Modern Miracles

Soon after the golden spike joined the nation's rails to create the first transcontinental railroad in May 1869, a second powerful industrial revolution roared in on the speed of assembly lines and the marvels of electricity. The second surge would also last a hundred years, and in that time more than three million new inventions would help to build the richest, most powerful nation on earth.

The second revolution began as a magnification of the first, but two world wars not only increased the flow of inventions but also the way they were created. Before, marvels like reapers and rifles, toothpicks and tractors, and peaveys and paper plates, came out of quite ordinary workshops. The products could be easily seen, examined, and understood, and often prompted users to wonder why they had not thought of making such things themselves. In the new age, explorations slowly moved out of workshops and into laboratories, and new discoveries were often beyond the scope of ordinary understanding.

Charles Best (1924).

Many of the new developments were in medicine, and none was more impactful than the discovery of a way to make insulin. Charles Herbert Best (1899-1978) of West Pembroke was a twenty-two-year-old medical student at the University of Toronto in 1921 when he and his mentor, Frederick Banting, were able to isolate the pancreatic hormone for the treatment of diabetes. It was a breakthrough discovery that would come to save countless lives far into the future.

Banting received the Nobel Prize for his work in 1924, but the committee declined to share the prize with Best because its rules forbade the honoring of a student. Banting did not agree, and promptly shared both the credit and the monetary prize with his young assistant.

William T. Bovie

William T. Bovie (1882-1958) was born and schooled in Michigan, but was closely tied to Maine. Trained as a botanist, he became fascinated with physics, and his work with radium and electricity was groundbreaking. As a research fellow at Harvard University, he was among the first to use radiation treatment for cancer, and it was his work with electricity that led him to develop the Bovie knife, a coagulating surgical generator first used in 1926 by the pioneering neurosurgeon Harvey Cushing. Modern versions of the eponymously named Bovie knife are still found in surgical suites throughout the world.

Bovie sold his rights to the invention to Harvard for a dollar and moved on to create the combined study of biology and physics, establishing the first biophysics program at Northeastern University in 1927. Two years later, Bovie moved to Bar Harbor to work at the new Jackson Laboratory and soon took an appointment as a physics lecturer at Colby College. He taught for ten years until his retirement in 1948, and lived another decade until his death in Fairfield on January 1, 1958.

William A. Rogers

Thirty years before Bovie's invention, another Colby physicist utilized science and electronics to establish precision in measurements. Working in a Waterville campus laboratory built especially for him, in 1890 William A. Rogers (1832-1898) measured wavelengths of sodium light to calculate the standard yard measure (yardstick) for the U.S. Bureau of Standards.

Joseph Stearns

Although Benjamin Franklin was experimenting with electricity in the mid-1700s, it was nearly a century after he stood in a thunderstorm with a metal key on a kite string that Samuel Morse invented the telegraph, and another forty years before Alexander Graham Bell came up with the telephone. After that, inventors began improving the system of telegraphic communication almost as fast as phenomenon itself.

Dozens of Maine tinkerers rode on that great wave, beginning with Joseph Stearns (1831-1895) of Weld, who was working as a telegraph operator for the Boston Fire Alarm Company in 1868 when he developed and patented the duplex system of telegraphy that allowed simultaneous, two-way communication. It changed the exploding world of electronic communication almost overnight, and he went on to lay miles of telegraphic cables in countries around the world.

Neville Hopkins

After the turn of the century, Neville Monroe Hopkins (1873-1945), a Portland native, took three degrees from Columbia University and then worked briefly for General Electric before forming his own consulting firm. On the eve of World War I, he was given charge of all U.S. Navy power plant installations, and then was called to help design projects for the U.S. Office of Public Roads. During the war, he volunteered with the French Red Cross and followed the French army on its retreat to defend Paris.

Following the Armistice, Hopkins wrote a book, *The Outlook for Research & Invention*, in which he exhorted American inventors to get back to work. "The shock of the great war just closing has, without the slightest doubt, been one of the greatest awakening influences our country has ever experienced," he wrote.

Taking his own advice, he proceeded to invent an array of military equipment, including long-range anti-aircraft and torpedo rockets, missiles, and ammunition. On the side of peace, he devised a synchronized electric registration and voting system that allowed for mass voting by means of radio communication. As an offshoot, he found a method of identifying the number of radio sets tuned to a particular station, an electronic system still used by radio and television rating services.

Henry Beverage

For Hopkins, growing up along the Maine coast was an extra inspiration in the exploration of transatlantic telegraphic communication. That certainly was true for one of his contemporaries who hailed from the tiny offshore island of North Haven. Harold Henry Beverage (1893-1993) was a nineteen-year-old electrical engineering student at the University of Maine when, on April 15, 1912, he intercepted the radio distress calls from the *RMS Titanic*.

Upon graduation, Beverage went to work for the General Electric Company, where he proceeded to invent the wave antenna, a landmark improvement making it possible to receive multiple shortwave messages, sort them out, and relay them over landlines.

During the war he operated the U.S. Navy receiving station at Otter Cliff in Bar Harbor, where he installed one of the first motorized, high-frequency radio transmitters, and at war's end he was sent to France to prepare a radio communication system for President Woodrow Wilson to use at the peace conference at Versailles. He later served nine years as chief engineer of the Radio Corporation of America (RCA), during which time he received more than twenty antenna patents, oversaw assembly of the first RCA television set, and designed and installed the first television antenna for the Empire State Building.

James A. Johnson

James A. Johnson was an inventor, author, race driver, and musician born in Portland. After graduating from Portland High School in 1928, Johnson enrolled at the Franklin Institute, a technical school in Boston. He served in the US Navy during World War II during which, amongst other things, he instructed ordnance trainees. He was given a special award by the United States for inventing a device to fire forty-millimeter guns by air, which helped improve safety. One of his loves was automobiles. Following his service, he invented an automotive exhaust emission device, a helicopter toy, a golf putter, and a wheel balancer fore cars and trucks. His Trig Stato-DiNamic Wheel Balancer for cars invented in 1959 was so revolutionary that it cut the costs to one sixth of that charged by the electronic wheel balancers that were then in use.

Johnson also helped establish Maine Vocational Technical Institute in 1946 in Augusta, in part to help train returning veterans. The school eventually moved and became what is now Southern Maine Community College. While an instructor at the school he wrote the popular text book, *Automotive Tune-up and Diagnosis.*

In 1974, Johnson was named Grand Knight of the Msgr. Houlihan Council in Portland, becoming the first black man to head a Knights of Columbus group in Maine.

Edward Norton

Edward Lawry Norton (1898-1983) was born not far from Henry Beverage in the mainland city of Rockland, and like Beverage, was also a communications engineer and scientist. A graduate of Massachusetts Institute of Technology, he joined the Western Electric Company (later Bell Labs) in 1922, at a time when the

Edward Lawry Norton

firm was racing to build a national telephone system. His most enduring contribution to the field was the development of an arcane yet useful means by which linear circuits are simplified by the use of a single current, thereafter known as Norton's Theorem. In his lifetime, he came to hold nearly twenty patents related to network theory and data transmission.

Percy Spencer

Another Maine contributor to the new age was physicist Percy Lebaron Spencer (1894-1970) of Howland. A leader in the design of radar tubes, he was building magnetrons in his laboratory at M.I.T. when he noticed the candy bars in his pockets were melting. He experimented in the laboratory with other food (including popcorn), and in 1945 filed a patent for the "Radarange," which two years later became the first commercial microwave oven.

The astonishing world of computers was cautiously welcomed and hard to comprehend. Bell Telephone created the transistor in 1947, and IBM made the first programmable computer in 1953. Five years later came the first integrated circuit (the chip), and the world moved into the high-speed information age in a single generation.

Charles Peddle

Maine's most notable contributor to the computer age was Charles "Chuck" Peddle (1937-2019), widely revered as the "father" of the personal computer. Born in Bangor, he grew up in Augusta in a poor family of six children, and with the help of scholarships and jobs from those who saw his promise, was able to attend the University of Maine's

Commodore PET 2001 computer, which used Peddle's MOS 6502 microprocessor.

Chuck Peddle
Courtesy of 1960 Prism *yearbook, University of Maine.*

engineering school, where he was among the first to concentrate in physics.

Upon graduation, Peddle took his first job with General Electric in California where he joined a research team working in computer design. G.E. abandoned computers in 1974, and he moved on to Motorola in Phoenix, Arizona, where he began work on a new, low-cost silicon chip that could be used for affordable personal computers. Motorola wasn't interested, and Peddle took his research team and the project to rival chipmaker MOS Technology in Pennsylvania where they built the acclaimed 6502 microprocessor, priced at $25 apiece. The invention was quickly adopted for use in the Apple I and II (1977), the Commodore 64 (1982), and game machines Acorn, Atari, and Nintendo, launching the personal computer revolution.

Paul André Albert

Another Maine computer pioneer was Paul André Albert (1926-2019) of Van Buren. Also a University of Maine engineering graduate, Albert first worked at Westinghouse and then went on to an inventing career at IBM. A metallurgist, he became a specialist in creating custom alloys used to record data on magnetic films for computer hard drives. When he retired in 1982, he established ACI Alloys, a firm that still provides custom alloy applications for scientists and manufacturers.

Epilogue

Most of the inventions that spurred the Industrial Age were promptly put on full display, and the common reaction of those who saw them was one of both head nodding in delight and surprise at not having figured out such things themselves. The same was true in the Machine Age that followed, when new things for homes, farms, and factories tumbled out of the nation's workshops at record speed. It wasn't until the end of World War II and the Atomic Age arrived that new creations began to slip beyond the grasp of ordinary understanding.

Everyone knew what horrible things an atomic bomb could do, of course, but few understood what made it work, and the gap in comprehension grew exponentially with the advent of the Computer Age in the 1980s. Everyone was astonished by what computers could do, but most had no idea how it was done. Today's evolution into a new age of artificial intelligence, robotics, and cyber-physical wonders has only deepened the gulf of incomprehension, yet despite the mysteries, the world has rushed to the small black boxes, keyboards and clickers that give out a flood of information. With the help of small children, even confirmed Luddites have surrendered to the digital age for fear of being left behind, and while many old industries have drifted away, the kinds of work that people do keeps changing, and the products of the nation slowly shift toward services, information, and ideas.

Today, new inventions and discoveries are most often made in exotic laboratories and the bellies of computers, and U.S. patents grow at the rate of nearly four hundred thousand a year. The Maine footprint in this age is as big as ever, albeit more centered at the growing number of scientific research centers around the state.

Few such centers are as prominent as Jackson Laboratory in Bar Harbor, where the discoveries of genetic solutions to human diseases save more lives every year, and where its sharing of genetically engineered mice helped scientists around the world in their search for therapies to stem the COVID-19 pandemic that began in 2020. Similarly, Abbott Laboratories in Scarborough became a national leader in developing COVID tests, and the nimble and versatile IDEXX Laboratories in Westbrook, a leader in animal diagnostics, swiftly responded to Maine's need for COVID testing by finding a way to provide enough kits to fill the state's urgent need.

There are other Maine inventors and discoverers at work at the Bigelow Laboratories for Ocean Sciences in East Boothbay, where scientists search for ways to save the world from its own mistakes, and at the Gulf of Maine Research Institute in Portland, where new ways are being found to protect the earth's ocean ecosystems.

Even as scientists make miracles in these and other Maine laboratories, there is nostalgia for the kind of homegrown inventions that emerged from the nineteenth century. It would be a mistake to think those days are over, especially in a state where staying at home is quite acceptable, and where the percentage of self-employed people ranks highest in the land. Beyond that, we know Edison's formula still works in Maine, where the spirit of curiosity and imagination never dies and where frugal people still revere a pile of junk. With the stars aligned, it seems certain Maine people will continue to join the world's unending quest to build a better mousetrap.

Acknowledgments

The idea of writing this book came from doing research for *Water Village*, a history of Waterville, when I ran across several remarkable Maine inventors who made significant contributions to the Industrial Revolution. I quickly discovered there were far too many to include in a comfortable book, and so I have limited the stories to those men and women whose clever creations clearly changed the world. If I have overlooked an obvious choice or two, forgive me.

I had much help along the way, but none more precious than that of Genevieve Morgan, the talented editor-at-large at Islandport Press, whose sharp eye and sound advice made this a better book. At Islandport, I also want to thank publisher Dean Lunt, editorial assistant Piper Wilber, and art director Teresa Lagrange, whose invaluable skills are apparent and much appreciated. I am also grateful for the gracious support of Bernard Fishman, director of the Maine State Museum, as well as Benjamin Stickney, the museum's curator of photography.

And, of course, I must thank my wife Barbara, who patiently read every word, and then read them again.

Several sources helped inform and illuminate these tales, especially the books listed in the bibliography as well as the Maine State Museum in Augusta, a rich and precious trove of fascinating Maine things. I also relied upon a number of reputable websites, including those of the U.S. Patent and Trademark Office, the New England Historical Society, National Geographic, Scientific American, the Smithsonian Institution, the International Directory of Company Histories (St. James Press), and of the digital archives of relevant manufacturers including J.I. Case, McCormick, International Harvester, Peavey Mfg., Huhtamaki, Goodyear, and L.L. Bean.

—Earl Smith

Bibliography

American National Biography, Oxford University Press, 1999

Eminent Mainers, Arthur Douglas Stover, Tilbury House, 2006

Hidden History of Maine, Harry Gratwick, The History Press, 2010

A History of the Keyes Fibre Company, Ernest C. Marriner, Colby College Press, 1958

Maine: A History, Louis Hatch, BiblioLife, 2009

Maine: A Narrative History, Neil Rolde, Tilbury House, 1990

Maine's Visible Black History, H. H. Price and Gerald E. Talbot, Tilbury House, 2006

Mothers and Daughters of Invention, Autumn Stanley, Rutgers University Press, 1994

My Story: The Autobiography of a Down East Merchant, L.L. Bean, 1960

Obed Hussey, Who of All Inventors, Made Bread Cheap, 1912, Project Gutenberg, 2006

The Outlook for Research & Invention, 1923, Nevil Hopkins, Reprint by BiblioBazaar

The Wild, Wild East, William Lemke, Yankee Books, 1990